AF352211

62

Advances in Biochemical Engineering/Biotechnology

Managing Editor: T. Scheper

Springer

Berlin
Heidelberg
New York
Barcelona
Budapest
Hong Kong
London
Milan
Paris
Singapore
Tokyo

Apoptosis

Volume Editor: M. Al-Rubeai

With contributions by
M. Al-Rubeai, F. Autuori, E. M. Bruckheimer,
S. H. Cho, T. G. Cotter, Z. Darzynkiewicz,
M. G. Farrace, N. L. Harvey, J. Herrmann,
S. Kumar, T. J. McDonnell, A. J. McGowan,
S. L. McKenna, R. O'Connor, S. Oliverio,
M. Piacentini, L. Piredda, M. Sarkiss,
R. P. Singh, F. Traganos

Springer

Advances in Biochemical Engineering/Biotechnology reviews actual trends in modern biotechnology. Its aim is to cover all aspects of this interdisciplinary technology where knowledge, methods and expertise are required for chemistry, biochemistry, microbiology, genetics, chemical engineering and computer science. Special volumes are dedicated to selected topics in which the interdisciplinary interactions of this technology are reflected. New biotechnological products and new processes for synthesizing and purifying these products are at the center of interest. New discoveries and applications are discussed.

In general, special volumes are edited by well known guest editors. The managing editor and publisher will however always be pleased to receive suggestions and supplementary information. Manuscripts are accepted in English.

In references Advances in Biochemical Engineering/Biotechnology is abbreviated as Adv. Biochem. Engin./Biotechnol. as a journal.

ISSN 0724-6145
ISBN 3-540-64153-X
Springer-Verlag Berlin Heidelberg New York

Library of Congress Catalog Card Number 72-152360

Typesetting: Fotosatz-Service Köhler OHG, Würzburg
Cover: Design & Production, Heidelberg
SPIN: 10573592   02/3020 – 5 4 3 2 1 0 – Printed on acid-free paper

Prof. Dr. S. B. Primrose
21 Amersham Road
High Wycombe
Bucks HP13 6QS/UK

Prof. Dr. P. L. Rogers
Department of Biotechnology
Faculty of Life Sciences
The University of New South Wales
Sydney 2052/Australia
E-mail: p.rogers@unsw.edu.au

Prof. Dr. K. Schügerl
Institute of Technical Chemistry
University of Hannover
Callinstraße 3,
D-30167 Hannover/FRG
E-mail: schuegerl@mbox.iftc.uni-hannover.de

Dr. K. Venkat
Phyton Incorporation
125 Langmuir Lab.
95 Brown Road
Ithaca, NY 14850-1257/USA
E-mail: venkat@clarityconnect.com

Prof. Dr. U. von Stockar
Laboratoire de Génie Chimique et
Biologique (LGCB)
Départment de Chimie
Swiss Federal Institute
of Technology Lausanne
CH-1015 Lausanne/Switzerland
E-mail: stockar@igc.dc.epfl.ch

Prof. Dr. H. J. Rehm
Institute of Microbiology
Westfälische Wilhelms-Universität Münster
Correnstr. 3, D-48149 Münster/FRG

Prof. Dr. H. Sahm
Institute of Biotechnolgy
Forschungszentrum Jülich GmbH
D-52425 Jülich/FRG
E-mail: h.sahm@kfa-juelich.de

Prof. Dr. G. T. Tsao
Director
Lab. of Renewable Resources Eng.
A. A. Potter Eng. Center
Purdue University
West Lafayette, IN 47907/USA
E-mail: tsaogt@ecn.purdue.edu

Prof. Dr. J. Villadsen
Department of Biotechnology
Technical University of Denmark
Bygning 223
DK-2800 Lyngby/Denmark

Prof. Dr. C. Wandrey
Institute of Biotechnology
Forschungszentrum Jülich GmbH
D-52425 Jülich/FRG
E-mail: c.wandrey@fz-juelich.de

Attention all "Enzyme Handbook" Users:

Information on this handbook can be found via the internet at

http://www.springer.de/chem/samsup/enzym-hb/ehb_home.html

At no charge you can download the complete volume indexes Volumes 1 through 13 from the Springer www server at the above mentioned URL. Just click on the volume you are interested in and receive the list of enzymes according to their EC-numbers.

Preface

Apoptosis is one of the two mechanisms by which cells die. Its role has been widely recognised in many of the industrialised world's major diseases such as cancer, immune system and neurodegenrative disorders and heart disease. The importance of apoptosis in cell culture and biopharmaceutical production has also been widely recognised and the goal of discerning its regulation is worth pursuing. The intellectual challenge and rewards are enormous, and are reflected in the 8100 papers on apoptosis published in the past three years.

Apoptosis (programmed cell death) is now defined as "a single deletion of scattered cells by fragmentation into membrane-bound particles which are phagocytosed by other cells" (Stedman's Medical Dictionary, 1995). The morphological features of apoptosis are quite distinct from necrotic cell death which typically occurs when cells are exposed to such severe stress that cell survival become impossible. These features include condensation of nuclear chromatin, fragmentation of DNA into oligonucleosomal fragments, cell shrinkage and the formation of membrane-bound "apoptotic bodies".

The chapters in this volume are intended to review the state-of-the-art with in-depth assessments of apoptosis. The aim of the volume is to make the recent developments in apoptotic research readily accessible to biotechnologists and biochemical engineers. The implication of apoptosis in the suppression of cancer and viral infection is presented to indicate the great potential of apoptotic research for the development of human therapies.

In this volume the characteristics and significance of apoptosis are highlighted. Biochemical and morphological changes during apoptosis and the methods used to detect these changes are described to give the reader useful information on how to use techniques such as fluorescence microscopy and flow cytometry to obtain quantitative data on apoptosis. A large part of this volume focuses on the genetic regulation and mechanisms of apoptosis, with emphasis on the identification of the key molecular sensors, mediators and modulators in apoptotic pathways and characterisation of their roles. Throughout the chapters a wide variety of both physiological and non-physiological agents that can induce apoptosis are revealed. Conversely, the role of survival factors in suppression of apoptosis is illustrated in a separate chapter with emphasis on the role of insulin-like growth factor and its receptor. The chapter on tissue transglutaminase describes the role of this effector element in the death pathway. The last chapter is dedicated to highlighting the progress and opportunities in the field of cell culture engineering. The strategies that have been undertaken to prevent the induction of apoptosis in cell culture and those which have been suggested as possibilities to improve culture productivity through the apoptosis route are discussed with given examples.

Birmingham, UK   Mohamed Al-Rubeai
February 1998

Contents

Molecular Mechanisms of Programmed Cell Death

Sharon L. McKenna · Adrian J. McGowan · Thomas G. Cotter
Tumour Biology Laboratory, Department of Biochemistry, University College, Lee Maltings,
Prospect Row, Cork, Ireland, *E-mail: stbi8033@bureau.ucc.ie*

Programmed cell death and apoptosis have now been recognised as biological phenomena which are of fundamental importance to the integrity of organisms. What may have evolved as an altruistic defence against pathogen invasion in simple organisms is now a major regulatory mechanism in the development and maintenance of multi-cellular organisms. The classically defined morphological characteristics of apoptosis are now accompanied by a plethora of information regarding common biochemical and genetic mediators of programmed cell death. It is apparent that life and death decisions are taken by individual cells based on their interpretation of physiological signals, or their own self-assessment of internal damage. The knowledge that cell death is a genetically regulated process has highlighted an inherent potential for manipulation and offered new avenues for research into several diseases, and also productivity improvements in the biotechnology industry. This relatively "new frontier" in cell science has undoubtedly widened our perspectives and may provide novel strategies to expedite both medical and biotechnological research.

Keywords: Apoptosis, Programmed cell death, Homeostasis, DNA damage, Genes, Viruses, Biotechnology, Disease.

Advances in Biochemical Engineering /
Biotechnology, Vol. 62
Managing Editor: Th. Scheper
© Springer-Verlag Berlin Heidelberg 1998

List of Symbols and Abbreviations

CML	chronic myeloid leukaemia
$\Delta\Psi_m$	mitochondrial transmembrane potential
GST	glutathione *S*-transferases
LRP	lung resistance protein
PARP	poly(ADP-ribose) polymerase
PCD	programmed cell death
PDGF	platelet derived growth factor
ROS	reactive oxygen species
VM26	Tenoposide

1
Introduction

Recent research has demonstrated that individual cells from multi-cellular organisms and several unicellular organisms have evolved biological mechanisms which grant them control over their own demise. Physiological cell death is a genetically controlled, internally mediated response to external stimuli. The knowledge that death is programmed from within brings with it the important conceptual advance that, except in the case of acute pathological cell death, external stimuli are not the inducers of cell death. Rather it is the cellular response to the stimulus that determines life or death. The discovery of various cell death modulators in the past decade has made it possible to manipulate the capacity of cells to initiate a program for cell death. This advance has important implications for the biotechnology industry where cell survival in adverse conditions can significantly enhance productivity. In medical research an ability to manipulate the threshold for cell death could be of significant value in the treatment of degenerative diseases which are characterised by excessive cell death, or in chemoresistant cancers where cell death is difficult to induce.

The intention of this Chapter is to introduce the concepts of programmed cell death to unfamiliar readers by giving a brief historical background on its discovery and surmising reports of its incidence. Morphological and biochemical features of physiological cell death will then be discussed, followed by an overview of cellular responses to some typical death inducing stimuli. A review of the genetic regulation of cell death and its exploitation by viruses will provide additional background and a basic introduction to subsequent detailed Chapters on specific gene families. Finally we have attempted to pre-empt discussions on the growing potential for exploitation of cell death mechanisms.

2
A Historical Overview of Cell Death

Physiological induction of cell death was recognised by embryologists as early as the 19th century. The formation of complex structures such as limbs, the disappearance of evolutionary vestigial structures and sexual differentiation were found to involve a predictable and concise amount of cellular destruction. The large-scale elimination of tissue that accompanies metamorphosis in insects and amphibia, and the formation of plant structures such as wood, were further examples of physiologically initiated cell death [1]. In 1964 Lockshin and Williams introduced the term *programmed cell death* (PCD) to describe the cellular demise which follows a predetermined sequence of events in development. Additional studies demonstrated that PCD in specific cell types is not necessarily inevitable, and can be influenced by physiological signals such as hormones from other cells [2, 3]. Further experiments using inhibitors of RNA and protein synthesis highlighted the importance of macromolecular synthesis in developmental cell death and fostered the concept of an intrinsic cellular control mechanism [4–6].

In what is now regarded as a landmark paper, Kerr et al. [7] presented evidence for a morphologically distinct form of cell death which they referred to as *apoptosis*. Structural changes in apoptotic cells included nuclear and cyto-plasmic condensation followed by dissociation of the cell into membrane bound fragments. These common structural features implied the presence of an active, inherently programmed phenomenon. A clear distinction could be drawn between apoptotic cell death (which included many incidences of PCD and death induced by a variety of physiological and pathological stimuli), and *necrotic* cell death (induced by acute pathological stimuli) which is a passive cellular degeneration resulting in cell lysis and concomitant local inflammation [7, 8].

As well as defining a morphologically distinct form of cell death, Kerr et al. [7] further suggested that the presence of apoptotic cells in normal adult tissues indicated that it represented a normal homeostatic mechanism, which is com-plementary to mitosis in the regulation of cell populations. Although this conceptual advance somewhat elevated the biological significance of apoptosis, its importance was subsequently relegated far below that of proliferation and differentiation in the early 1980s. The reasons for this are probably numerate; an innate reluctance to find anything positive or progressive about cell death was probably a significant contributory factor! Undoubtedly the lack of genetic evidence for PCD added to the scepticism. Although the concept of an intrinsi-cally controlled mechanism of cell death implied genetic control, the absence of any known genetic regulators for PCD rendered this a tentative assumption.

Direct evidence for a genetically controlled cell-death pathway came from stu-dies of developmental mutants of the nematode worm *Caenorhabditis elegans*. Of the 1090 somatic cells formed during the development of this hermaphrodite, 131 are destined to die, displaying morphological features which resemble apoptosis. Mutants defective in this developmental cell death provided powerful tools for genetic analysis. Two genes, *CED-3* and *CED-4*, were found to be required for nor-mal developmental cell death. Another gene, *CED-9*, was identified as a negative regulator of cell death and acts antagonistically to *CED-3* and *CED-4* to suppress cell death. Other genes were shown to mediate the engulfment and degradation of apoptotic corpses [9, 10]. These studies set a precedent for the acceptance of apoptosis as a genetically directed program of self-destruction, and established apoptosis as a positive cellular response analogous to proliferation and differen-tiation [11]. The identification of human homologues for both *CED-9* and *CED-3* gene products propelled apoptosis into mainstream research [12, 13]. Many new genes and previously identified genes have now been shown to influence cell death, and indeed large gene families have been identified whose members appear to be almost exclusively involved in the regulation or execution of cell death [14, 15] (see also Chaps. 3, 4 and 5 this volume). In a relatively short period of time, the pace of apoptosis research has increased such that the literature on the subject accumulates at a rate of ~ 200 papers per month.

At this point it may be useful to comment on a matter of terminology. PCD and apoptosis are terms which are often used synonymously, much to the irritation of some pioneers in the field! [16, 17]. As the term *programmed cell death* was initially used to describe pre-determined deaths during development, it incor-porates a functional and temporal aspect into the induction of cell death. It does not discriminate on any morphological or mechanistic aspects of cell death. However, as knowledge of both the incidence and importance of physiological cell death has expanded, PCD tends not to be confined to development and is now generally used to refer to any cell death that is mediated by an intracellular death program regardless of the initial trigger [18]. Apoptosis is a descriptive term which defines a set of morphological (and now also biochemical) features (discussed in detail below). While many PCDs share the morphological features of apoptosis, there are exceptions which do not, particularly in plants and invertebrates. It remains to be established whether or not these are completely different mechanisms of cell death or different manifestations of a similar internal programme.

3
Incidence of PCD and Apoptosis

Since its "rise to prominence" PCD has been identified in every multi-cellular organism in which investigators have looked for it. The morphological and biochemical features of apoptosis have been observed in a wide range of species, suggesting an early evolutionary origin and a fundamental biological significance. The extent of molecular conservation is such that a human gene (*Bcl-2*), can functionally compensate for the absence of its homologue (*CED-9*) in an invertebrate [19]. Although the origins of PCD were initially associated with the attainment of multi-cellularity, a number of unicellular organisms can also initiate their own destruction and may exhibit morphological features resembling apoptosis [20, 21]. The incidence and importance of PCD/apoptosis can be exemplified by considering several functional categories.

3.1
Viral/Bacterial Defence Mechanism

It has been suggested that the most likely explanation for the evolution and conservation of a program for a cell death is that altruistic cell death is a potent defence against infection by viruses [22]. This anti-viral strategy is largely substantiated by the by the fact that many successful viruses carry anti-apoptotic genes (discussed in detail in Sect. 7). An anti-viral suicide response mechanism exists in certain strains of *E. coli* [23, 24] and in the hypersensitivity response of plants, the detection of a pathogen by a single cell initiates a suicide response in that cell, and in other cells in its immediate vicinity [25]. It is possible that the activation of endonucleases early in apopotosis (see Sects. 4.1 and 4.2) evolved as a mechanism for eliminating viral DNA. Preservation of this mechanism prevents the transfer of intact potentially deleterious genetic material when apoptotic bodies are phagocytosed by neighbouring cells [26].

3.2
Developmental Significance

As mentioned in the previous section, PCD was first identified in developmental studies. The controlled deletion of cells plays a crucial role in the removal of unnecessary tissue in the sculpting of limbs and organs. A well quoted example is the formation of digits by the programmed cell death of inter-digital tissue [27]. Similarly, the PCD of internal cells in solid structures facilitates the formation of lumina [28]. The importance of PCD has also been noted in palatal fusion [29], remodelling of cartilage and bones [30, 31], formation of the retina [32], and in the development of the nervous system [33, 34].

While the inhibition of developmental cell death is non-lethal in simple organisms such as *C. elegans* [11], it is undoubtedly indispensable in higher organisms. PCD-deficient flies die early in development [35] and CPP32 (cell death mediator; see Sect. 6.7) knock-out mice die in development with a vast excess of cells in their central nervous system [36].

3.3
Homeostatic and Specialised Functions in the Adult

In all normal proliferating adult tissues, mitosis is counterbalanced by apoptosis. Both inherent factors (such as differentiation status) and external factors (hormones, cytokines, extracellular matrix) influence cellular viability. Accumulating evidence suggests that all cells (with the zygote or blastomere being the only known exceptions) express a default program for cell death and depend upon survival signals from other cells for their viability [37]. This mechanism would restrict specific cell types to the tissue producing the required set of survival factors and limit the population size according to their availability (see also Chap. 6 this issue). The maintenance of functionally distinct populations of cells within the haematopoietic system is a classic example of population control largely mediated by soluble survival factors [38]. The haematopoietic system also employs mechanisms for selectively inducing apoptosis in functionally inadequate, potentially dangerous cells. For example, auto-reactive immature T-cells are negatively selected in the thymus and undergo apoptosis. Cytotoxic T-cells have developed elaborate independent mechanisms for inducing apoptosis in host cells harbouring intracellular pathogens, or in abnormal cells such as cancerous cells [39] (and Chap. 3 this issue).

3.4
Response to Injury or Stress

Excessive doses of a variety of non-physiological agents (e.g. UV irradiation, hyperthermia, metabolic poisons etc.) have been shown to induce necrotic cell death. Mild insults however, such as those which may be more realistic in the environment (e.g. carcinogens in the gut, skin exposure to UV) have been shown to induce apoptosis (see Sect. 5). Mechanisms which enable the recognition and physiological elimination of damaged and potentially dangerous cells are imperative for the survival of the organism.

4
Morphological and Biochemical Features of Apoptotic Cell Death

The stereotypical pattern of morphological events which have been described in the majority of apoptotic cells includes cell shrinkage, membrane blebbing, nuclear condensation and fragmentation, and finally the formation of sealed vesicles called apoptotic bodies [7] (see Fig 1). These apoptotic bodies ensure membrane integrity is maintained prior to phagocytosis. This scheme of events is in stark contrast with the cell swelling, membrane lysis and inevitable inflammatory response observed during necrosis [8].

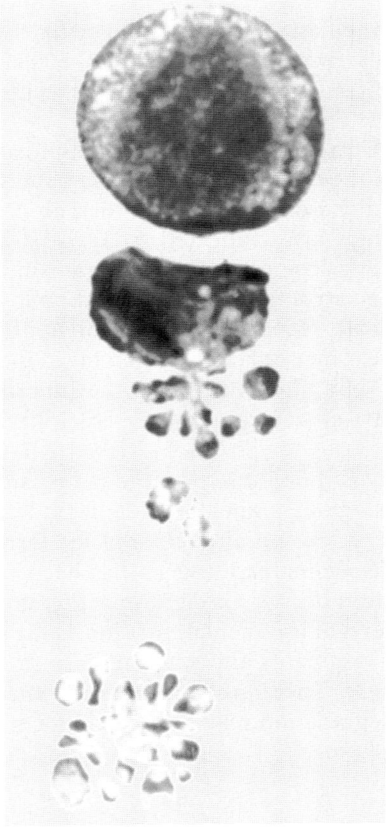

Fig. 1. Morphological features of apoptosis. Three cells in different stages of apoptosis. HL60 (myeloid leukaemia) cells were induced to undergo apoptosis by exposure to ultraviolet light. Cytocentrifuged cells were stained with Leukostat (Fisher Scientific, Orangeburg, NY). The *upper cell* in the photograph displays normal morphology. The *cell below* shows typical apoptotic morphology. It is shrunken in appearance and is releasing its cellular components in membrane bound fragments (apoptotic bodies). The *lower part of the photograph* shows the membrane bound remnants of a cell which has undergone apoptosis. Under physiological circumstances apoptotic cells undergo the later phases of disintegration after ingestion by a phagocytic cell

4.1
Nuclear Events During Apoptosis

Unlike necrosis, which lacks any internal self-regulation, apoptosis displays an intrinsic set of controls [16, 40]. The nucleus during apoptosis becomes condensed and marginated whereas a necrotic nucleus appears swollen. A major biochemical event associated with apoptosis in a range of cell types is the degradation of DNA into 180–200 bp oligonucleosomal fragments. These fragments were initially detected using agarose gel electrophoresis and referred to as the "DNA ladder" [40]. The DNA ladder was considered to be the hallmark of apoptosis until it was subsequently demonstrated that some cell types do not display this pattern of DNA fragmentation [41, 42]. However, in the absence of oligonucleosomal fragmentation or low molecular weight fragmentation (LMW), higher molecular weight fragments 300–500 kbp (HMW) have been reported [43]. This was first demonstrated using the technique of pulsed field gel electrophoresis [44].

The patterns of DNA fragmentation during apoptosis appear to be related to the manner in which the chromatin is packaged and folded in the genome. 140 bp of DNA is initially wound around a histone core particle to form a nucleosome. Nucleosomes are connected by linker DNA which may range from 10–100 bp in length depending on the organism. The nucleosomes are then arranged into coiled structures or loops, 50 kbp in size, and subsequently packed into hexameric rosette structures of approximately 300 kbp in length [44, 45]. The appearance of HMW fragments in the absence of, or immediately preceding, LMW fragments suggests that the dismantling of DNA is dictated by its quaternary structure, and the availability or activity of specific endonucleases.

Endonuclease activity is thought to be largely mediated by pH and the presence of specific cations. In one model for DNA degradation it has been suggested that magnesium ions are initially required for formation of HMW fragments, but as calcium levels increase (which has been reported during apoptosis), LMW fragmentation occurs. Calcium may therefore act as a rheostat to control the degree and level of fragmentation. Zinc on the other hand can inhibit the formation of LMW fragments and some reports suggest it can also block HMW fragmentation [43]. Therefore, any putative enzyme suspected to be involved during DNA degradation may have to conform to the stringent ionic conditions in a given cell type.

4.2
Candidate Endonucleases

The search for "the endonuclease" responsible for DNA fragmentation during apoptosis was initially considered to be the "holy grail" of apoptosis research. A number of different groups have attempted to isolate and purify the endonuclease responsible for DNA fragmentation during apoptosis. DNase I, DNase II, NUC 18/cyclophillin and several novel endonucleases been proposed to be involved in DNA fragmentation in apoptotic cells [46–52].

Studies of isolated nuclei suggest the presence of at least two different types of endonuclease activity, one of which is tightly bound to the chromatin and one which has ready access to the intra-loop regions of the DNA [43]. However, one interesting proposal outlines how one enzyme, namely DNAse I, could potentially catalyse all of the different stages of DNA fragmentation based on its exquisite sensitivity to cation levels [43]. The specific ionic requirements of DNase I suggest that its various enzymatic activities may be sequentially activated during apoptosis. [53]. Firstly, the enzyme can create single strand breaks at specific sites in the DNA in the presence of magnesium ions, resulting in the formation of fragments of >300 kbp. The next phase begins as calcium levels increase and the chromatin may be cleaved into fragments of 300–50 kbp in size. Finally, at the highest calcium levels, oligonucleosomal fragmentation occurs and during this stage protease activity is also evident. Although this evidence suggests a DNase I-like enzyme is capable of generating the different facets of DNA fragmentation observed during apoptosis, it is possible that other novel endonucleases also fulfil the same ionic requirements. Another group have reported that a DNase II-like activity predominates in Chinese hamster ovary cells [47]. DNase II requires an acidic pH for optimal activity. In accordance with this, intracellular acidification has been detected in several cell types during apoptosis [47]. In addition, Hughes and Cidlowski have identified two novel nucleases which appear to fit a specific ionic profile during apoptosis [54].

It is possible that the endonucleases activated in apoptotic cells may be dependent upon the initiating signal, cell type and/or differentiation status. It is interesting that a recent study reported that completely different types of endonucleases were activated in apoptotic bone marrow progenitor cells, and their granulocytic descendants [55]. Thus the identity of the enzyme or enzymes responsible for DNA fragmentation during apoptosis remains controversial and the subject of intensive research.

4.3
Cytoplasmic Components of Apoptosis

The focus of attention, with regard to both apoptosis regulation and the dismantling machinery, has recently diverted from the nucleus to the cytoplasm [56, 57]. Evidence from work completed in enucleated cells and cell free systems (and genetic evidence, see Sect. 6), has suggested that critical components of the apoptotic program reside in the cytoplasm [58, 59]. The identification of several key mediators in apoptotic pathways has enticed investigators into the "ICE AGE" of apoptosis research.

4.3.1
The ICE Age

ICE like proteases (now referred to as caspases) are a highly conserved family of enzymes which are sequentially activated during apoptosis induced by a wide variety of both physiological and non physiological agents. Several proximal enzymes act as key signal transducing proteins in specific apoptotic pathways (e.g. FLICE see Sect. 6 on genetic regulation, and Chap. 5 in this volume). Other more distal acting enzymes are considered to play a purely destructive role and are instrumental in the breakdown of cellular components prior to packaging into apoptotic bodies.

The number of different substrates identified as targets for proteolytic degradation continues to grow [56, 57]. These protein substrates range from components of the cytoskeleton such as actin and fodrin, to nuclear proteins such as poly(ADP-ribosyl)polymerase/PARP and Lamin B1. It is likely that proteolytic cleavage of several key proteins involved in cellular maintenance and/or repair propagates the induction of cellular collapse which we now refer to as apoptosis [56]. The manner in which this cellular collapse is orchestrated remains to be established but several observations indicate how proteolysis slots into the developing picture of regulated cell death.

For example, poly (ADP-ribose) polymerase (PARP) is recognised as a sensor of DNA damage, but once cleavage of this enzyme occurs its activity diminishes, and consequently DNA damage proceeds unchecked. Lamin B cleavage may lead to the controlled collapse of chromatin structure evident during apoptosis. An interesting observation is that cleavage of actin abrogates its inhibition of DNase I, and thus may enable the enzyme to catalyse the fragmentation of DNA [60]. Furthermore, cleavage of actin-associated fodrin may accommodate the changes observed at the membrane level such as blebbing and apoptotic body formation [61].

4.3.2
Mitochondrial Control

In addition to the cytoplasmic-based caspases, another cellular component resident in the cytoplasm, the mitochondria, has been implicated in the controlled demise of an apoptotic cell [62, 63]. A number of reports have described a reduction in the mitochondrial transmembrane potential ($\Delta\Psi_m$) prior to any other of the conventional changes associated with apoptosis. This reduction in $\Delta\Psi_m$ has been demonstrated in a range of different cell types in response to a variety of insults. Furthermore, cells displaying a reduced $\Delta\Psi_m$ are irreversibly committed to die. Isolated mitochondria undergoing reductions in $\Delta\Psi_m$ were also found to induce nuclear alterations, typical of apoptosis, in a cell free system [63].

Arguments disputing mitochondrial participation in the regulation of apoptosis have arisen from work with cells which have been depleted of mitochondrial DNA, and yet can still be induced to die via apoptosis [64, 65]. However, more recent evidence has shown that cells depleted of mitochondrial DNA still display a reduced $\Delta\Psi_m$ during apoptosis. Therefore, mitochondrial functions critical for the induction of apoptosis are encoded by the nucleus [63]. Zamzami et al. also suggest mitochondria contain a soluble pro-apoptotic factor that mediates the changes associated with apoptotic cell death [63]. This factor is released following mitochondrial membrane depolarisation. The potent inhibitor of apoptosis, *bcl-2* (see Sect. 6.1 on genetic regulation), has also been suggested to work by controlling changes in $\Delta\Psi_m$, factor release and subsequent radical production.

A further interesting observation is that when the induction of thymocyte apoptosis by Fas was blocked by the use of a specific ICE inhibitor, a reduction in $\Delta\Psi_m$ was also simultaneously prevented. These observations appear to link proteases and mitochondria to the regulation of apoptotic cell death. In addition, it suggests that caspase activity and reduced $\Delta\Psi_m$ may be sequential events in a common pathway or at least closely interlinked pathways to cell death.

4.4
Cytoskeletal and Cell Membrane Alterations During Apoptosis

Calcium dependent transglutaminase catalyses the cross-linkage of proteins resulting in the formation of $\varepsilon(\gamma$-glutamyl) lysine linkages between polypeptides (See Chap. 7 for greater detail). This ensures membrane integrity and stability, thus limiting cellular leakage during apoptosis. Transglutaminases have also been implicated in apoptotic body formation [66].

Cell membrane blebbing is another characteristic feature of an apoptotic cell. Calcium elevation and re-distribution appear to play a role in the formation of membrane blebs. In addition, proteolytic cleavage of constituents of the cytoskeleton such as actin and fodrin may facilitate bleb formation [61]. The importance of maintaining cytoskeletal structure is evident from the disruptive effects of microtubule targeting drugs such colchicine or vincristine, both of which induce apoptosis. In contrast, the inhibition of actin polymerisation by cytochalasin B interferes with apoptotic body formation [56].

The major purpose behind many of the changes associated with apoptosis may be to make an apoptotic cell more readily identifiable and palatable to neighbouring cells or professional scavengers like macrophages. Changes at the cell surface during apoptosis have been linked to the recognition phase of apoptosis. The interaction between the vitronectin receptor $\alpha_v\beta_3$ integrin on human neutrophils, the CD36 ligand, present on macrophages, and thrombospondin, allows the formation of a bridge between the neutrophil and macrophage, thus ensuring recognition and phagocytosis of the dying cell [67].

Another cell membrane alteration commonly observed during apoptosis is the re-distribution of specific phospholipids in the cell membrane during apoptosis [68]. Normal cell membranes display a markedly distinct phospholipid distribution with phosphatidylcholine and sphingomyelin predominately localised to the external surface, whereas phosphatidylethanolamine and phosphatidylserine are on the inner surface. At the onset of apoptosis, loss of membrane phospholipid asymmetry results in the appearance of phosphatidylserine on the outer surface. This re-shuffling of membrane phospholipids, and in particular the external appearance of phosphatidylserine, is now a characteristic feature which can be used to detect apoptosis in a range of cell types using the phosphatidylserine specific probe annexin-V ([68] and Chap. 2 in this volume). Therefore, at least two distinct membrane associated signals have been identified which can facilitate the recognition and removal phase of apoptosis.

5
Mechanisms of Apoptosis Induction

While it has now been established that a broad range of both physiological and non-physiological stimuli can result in apoptotic cell death, relatively little is known about how cells interpret these signals and initiate a program for cell death. Evidence suggests that cells do not passively submit to non-physiological cytotoxic stimuli, but rather will undergo a considerable degree of damage limitation and self assessment before opting for death as a last resort. This necessitates the presence of various control points in the pathway. For example, the insult must be detected by what is probably a specific sensor. The extent of the damage must be assessed by either the sensor or a distal mediator and the appropriate response then initiated. In the case of sub-lethal damage, specific repair mechanisms would be initiated enabling cell rescue. Excessive damage must however be recognised as unrepairable and unsustainable, leaving death as the only available option (see Fig. 2).

5.1
Inducers of Apoptosis

5.1.1
Physiological Triggers: Cytokines, Growth Factors, Hormones and Intergins

Cytokines are low molecular weight signalling molecules which bind to high affinity receptors on target cells. They may provide signals for cell survival, proliferation, differentiation or death. Death inducing cytokines (e.g. FasL, TNF, see Sect. 6.6) can act as membrane bound ligands or soluble secreted factors. Fas and TNF receptor ligation induces apoptosis in the absence of RNA or protein synthesis [69, 70] and also in enucleated cells, suggesting that the components required for the induction of apoptosis are constitutively expressed in susceptible cells.

Several growth factors and hormones have been shown to promote cell survival. For example, neuronal cells depend upon nerve growth factor for their viability and breast and prostate cells are dependent upon oestrogen and androgens respectively ([71] and Chap. 6 this volume). Glucocorticoids are steroid hormones which are involved in the induction of apoptosis in immature T-cells in the thymus. The cell death induced in this instance requires protein synthesis [40]. Adhesion molecules (integrins) have also been shown to promote cell viability upon binding to specific cell surface receptors [72].

In contrast to the fairly well established sequence of events during Fas/TNF induced apoptosis (see Sect. 6.6 and Chap. 4 of this volume), relatively little is known about the mechanisms involved in the induction of apoptosis following survival factor withdrawal.

5.1.2
Non-Physiological Triggers of Apoptosis

One of the most important initial observations with respect to drug induced cell death was that etoposide (a topoisomerase II inhibitor) induced rapid inter-

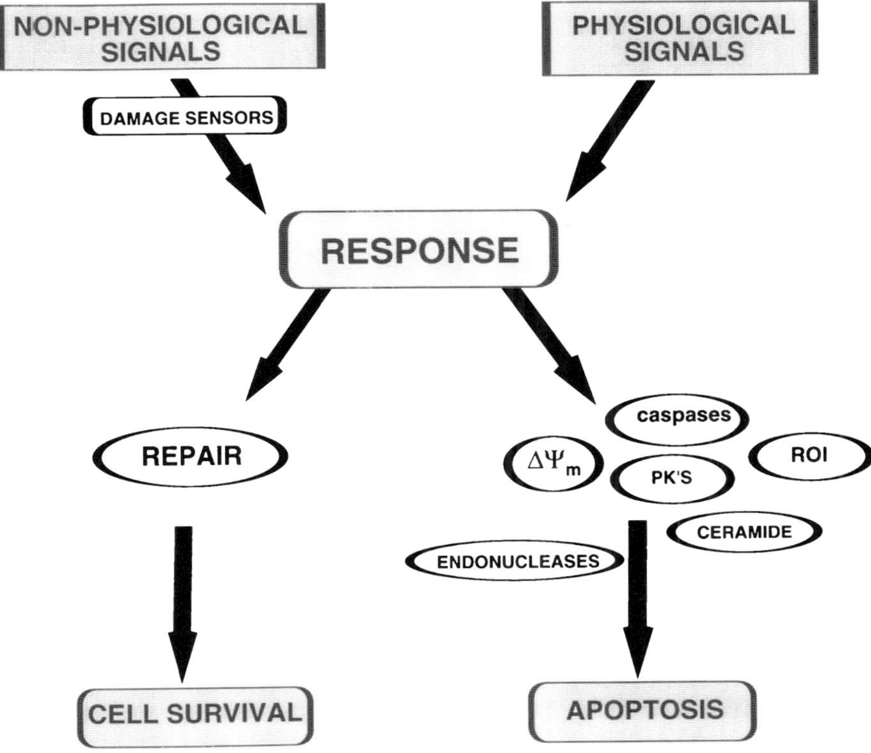

Fig. 2. Life and death decisions. The interpretation and response of a mammalian cell to both physiological and non-physiological stimuli is depicted here in schematic form. Physiological stimuli may be directly interpreted as signals for either cell survival (e.g. survival factors) or cell death (e.g. Fas ligation). Cellular damage which has been caused by non-physiological signals is initially assessed by damage sensors (e.g. PARP, p53). Depending on the extent of the damage the response may be cell cycle arrest and repair, or alternatively the induction of apoptosis. Known examples of intracellular signalling molecules have been included in the diagram but undoubtedly many unknown mediators are also involved

nucleosomal fragmentation [73]. This chemotherapeutic agent was thus recognised as a potent inducer of apoptosis. The spectrum of diverse acting cytotoxic agents known to trigger an apoptotic response has been progressively increasing, and now includes inhibitors of protein and RNA synthesis (e.g. cyclohexamide, actinomycin-D), dihydrofolate reductase inhibitors, topoisomerase I and II targeting drugs (e.g. camptothecin, etoposide, VM26, doxorubicin), nucleoside analogues (cytosine arabinoside, fludarabine and 2-chloro-2′-deoxyadenosine), microtubule poisons (vinblastine and vincristine), alkylating agents, cisplatin, ionising radiation, and hydrogen peroxide (reviewed in [74]). Many of these agents have been shown to provoke consistently an apoptotic response in a variety of cell types. These studies demonstrate that, although most cytotoxic agents differ in their primary targets, they inevitably converge to induce an apoptotic response. It is therefore possible that both physiological and non-physiological agents utilise common molecular mediators in the induction of apoptotic cell death.

5.2
Damage Limitation and Repair

While a detailed review of mechanisms of cellular resistance and repair is beyond the scope of this review, it is perhaps appropriate at this point to mention some of the better known strategies for damage limitation. For example cells may utilise energy driven pumps (e.g. P-glycoprotein, Multi-drug Resistance associated Protein, MRP) which will extrude cytotoxins from the cell against a concentration gradient, thereby effectively resulting in a decrease in the intra-cellular concentration of the toxin. Other cytoplasmic mechanisms can result in drug detoxification (e.g. GST enzymes) or in the sequestering of drugs away from the nucleus (e.g. LRP vaults). In the nucleus, drug target modifications may also alter cytotoxicity (e.g. down-regulated or mutated topoisomerase enzymes) (reviewed in [75]).

Cells also possess several mechanisms which enable them to repair damaged DNA. For example mis-match repair and base excision repair can deal with sub-lethal damage induced by ionising radiation such as UV and X-rays, and also DNA damage caused by alkylating agents and cisplatin [76, 77]. Eukaryotic cells can also induce a heat shock response following mild hyperthermia. The expression of a variety of evolutionary conserved heat shock proteins (hsps) protects cellular constituents from heat denaturation and can also afford a certain degree of protection against other cytotoxic stimuli [78]. Many of these protective mechanisms may be attenuated following a prior mild exposure, allowing the cell to tolerate levels of exposure which might previously have resulted in the induction of apoptosis.

A variety of enzymes also protect cells from oxidative stress. Reactive oxygen species (ROS) produced in normal aerobic respiration, or as a consequence of other injury, may be neutralised by anti-oxidant enzymes such as superoxide dismutases, catalase and glutathione peroxidases. Moreover, accumulating evidence suggests that the regulation of ROS is of more than homeostatic relevance, and that ROS may be crucial intra-cellular signalling molecules involved in the induction of apoptosis (Sect. 5.4.4).

5.3
Molecular Sensors of DNA Damage

5.3.1
p 53

Possibly the best known molecular sensor/responder following DNA damage is the *p53* tumour suppressor gene. This gene mediates a transient arrest of the cell cycle to allow time for damage assessment and DNA repair. If damage is excessive, p53 executes a program for cell death. p53 therefore prevents the replication of cells with damaged DNA. Several studies suggest that it may also mediate the induction of apoptosis following growth factor withdrawal ([79, 80] and Sect. 6.2 on genetic regulation).

5.3.2
Poly(ADP-ribose)polymerase/PARP

DNA strand breaks also trigger the rapid synthesis of poly(ADP-ribose) by the enzyme PARP [81]. This enzymatic response has provoked intensive investigation as PARP activation precedes p53 induction of DNA repair [82]. At the site of a DNA strand break, PARP catalyses the transfer of an ADP-ribose moiety from its substrate nicotinamide adenine dinucleotide (NAD) to a number of different protein targets involved in chromatin structure and DNA metabolism. Initially it was speculated that PARP was central to all DNA repair processes but there have been a number of reports now demonstrating a role for PARP in base excision repair but not nucleotide excision repair [81, 83]. More recently, PARP-deficient mice were developed and these animals were shown to exhibit efficient DNA repair responses. Thus other molecular sensors must be present and the exact role of PARP in the recognition and repair of DNA damage remains to be established.

5.4
Intra-Cellular Signalling Molecules

Given the extensive range of apoptosis inducers, it would perhaps be unsurprising to find a diverse range of signalling molecules feeding into a number of pathways to apoptosis. Various molecular mediators including protein kinases/phosphatases, tyrosine kinases, calcium, ceramide and oxygen radicals have been implicated in signalling cascades [84, 85]. Considering the degree of conservation in morphological and biochemical events at the execution stage, it is conceivable that all signals may converge upon a limited number of common final pathways [86, 87]. The existence of a limited number of final pathways may explain how *bcl-2*, and other negative regulators of apoptosis (see Sect. 6) can inhibit apoptosis which has been initiated by a broad spectrum of stimuli.

5.4.1
Kinases

Protein tyrosine kinases (PTK) transmit many of the signals from cell surface receptors to the nucleus [88]. A number of known inducers of apoptosis have been shown to utilise PTK cascades to signal apoptosis. Research has shown that ionising radiation requires PTK activation for B cell apoptosis [89]. In contrast, the CML cell line K562 displays increased tyrosine kinase activity, a feature which has been linked to its resistance to apoptosis [90]. In addition to tyrosine kinases, serine threonine kinases may also play a role in signal transduction. For example, p34^{cdc2} which controls entry into the cell cycle is dephosphorylated prior to the onset of apoptosis [91].

5.4.2
Protein Kinase C and Calcium

Surface receptors can also link up with Ca^{2+} signalling pathways via G proteins and PTKs which in turn activate phospholipases C and D to produce inositol triphosphate and diacylglycerol. Inositol triphosphate then prompts Ca^{2+} release from the ER and diacylglycerol activates the serine threonine kinase and protein kinase C (PKC). Calcium influx has, for some time now, been implicated as a trigger for apoptosis [85]. However, in the majority of cases little or no changes in cellular calcium levels appear to be essential for the induction of apoptosis.

5.4.3
Ceramide

Sphingomyelin, a membrane based lipid, appears to act as a damage sensor which can be tripped by a diverse range of insults, resulting in the formation of ceramide following sphingomyelinase activation [92–94]. Ceramide is then thought to act as a second messenger, mediating the induction of apoptosis. The insults which result in ceramide production include TNF, Fas activation, cross-linking of surface immunoglobulin receptors and ionising radiation [94]. A designated number of targets have been proposed to account for the diversity of its actions including a ceramide-activated protein kinase, ceramide-activated protein phosphatase and PKCζ (reviewed in [95]). A common converging point in many of these signalling pathways is the activation of transcription factors such as NF-κB and AP-1. Induction of apoptosis in HL-60 leukaemic cells by ceramide has been shown to require AP-1 [96]. In another report, a number of diverse acting cytotoxic agents were also shown to activate NF-κb prior to HL-60 cell death [97] and author's unpublished observations.

5.4.4
The Common Mediator Hypothesis

The ability of multiple signalling pathways to engage what appears to be a common pathway(s) to cell death suggests that a potential common mediator must link a diverse range of signals to the activation of downstream executioners (e. g. caspase family members; see Sect. 6.7. on caspases). Oxidative stress or changes in the cellular redox state has been proposed as such a common mediator [86, 87, 98]. There are several lines of evidence which support oxidative stress as a candidate mediator. First, the direct addition of oxidants such as hydrogen peroxide rapidly induces apoptosis in many cell types [99]. Second, many agents that are not themselves oxidants elicit reactive oxygen intermediate (ROI) production during the induction of apoptosis [98, 100, 101]. These agents may achieve this in several ways. They can deplete the cell of major cellular defence mechanisms which protect against oxidative stress, for example, glutathione. This has been demonstrated in Jurkat T-cells following Fas activation [87]. Alternatively, activation of an oxidant producing enzyme such as NADPH oxidase has been implicated in the hydrogen peroxide formation during amyloid β protein cytotoxicity [102]. Another possibility is the disruption of mitochondrial enzyme activity or transmembrane potential [63, 103]. Further evidence that oxidative stress is a common trigger for many cytotoxic agents is the efficacy with which many anti-oxidants prevent apoptosis following cytotoxic insult [98, 104, 105]. The deleterious effects of ROI production is underscored by the intricate network of intracellular anti-oxidant enzymes and molecules. Together this defensive arsenal functions to block the production (tocopherol, carotenoids and ascorbic acid), scavenge the products (superoxide dismutase, glutathione peroxidase, catalase and thioredoxin) or interfere with the catalysed generation of ROI [106].

However, there are arguments which contest the validity that oxidants may act as common mediators. The evidence against this hypothesis has come from experiments performed under near anaerobic conditions to reduce oxygen radical formation [107, 108]. However, these experiments under hypoxic conditions may be misleading as, under hypoxia, cells down-regulate important anti-oxidant enzymes and so the threshold at which a cell experiences an oxidative stress would also be lowered [87]. Furthermore, oxidative damage or changes in cellular redox status can proceed unabated in the absence of oxygen.

Results from several laboratories have implicated hydrogen peroxide as the ROI involved during the induction of apoptosis [98, 100]. Oxidative stress could potentially fulfil the role of mediator for many of the other signalling cascades as ROI production can promote tyrosine kinase activity, calcium release and ceramide production [93, 109]. The pluripotent effects of ROI production may then explain how these signalling pathways overlap and interconnect. Furthermore, it has been demonstrated that oxidatively modified proteins are more prone to proteolytic degradation [110]. This may link changes in cellular redox status to the activation of the caspase family as oxidative modification of an inactive pro-enzyme may target this zymogen for proteolytic activation, thus leading to a proteolytic cascade culminating in apoptosis [87]. Recent work from our laboratory has demonstrated a link between oxidative stress and the proteolytic cleavage of PARP [98]. The signal transduction phase of apoptosis remains to be fully determined but the potential for manipulating cell death from within may well reside in delineating the molecular events during this phase of apoptosis.

6
Genetic Regulation of Cell Death

The initial discovery of gene-directed cell death in *Caenorhabditis elegans* fuelled an intensive search for new genes exclusively involved in the regulation or execution of cell death pathways. Many new cell death regulators have been identified, and several gene families largely comprised of cell death mediators have emerged. In addition, a number of previously identified oncogenes or suppressor genes (e.g. *myc, ras* and *p53*), have been shown to modulate pathways for cell death. Functional and comparative studies with some cell death regulators have enabled a limited amount of molecular ordering in the cell death pathway (see Fig. 3), and have localised major control points to the cytoplasm.

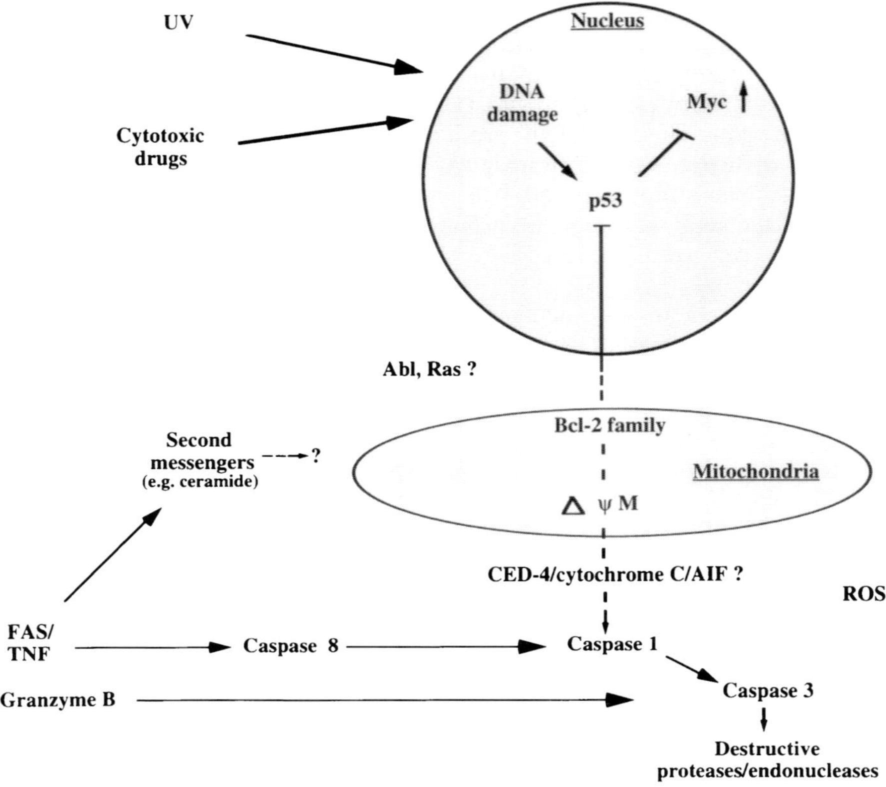

Fig. 3. Molecular ordering in the cell death pathway. Evidence from the use of various inhibitors or inducers of apoptosis has allowed a limited amount of molecular ordering in apoptosis pathways. This schematic diagram has attempted to illustrate some of the perceived ordering in molecular events. It is undoubtedly simplistic and by no means fully inclusive as evidence suggests that the same stimulus may feed into several apoptotic pathways. For example Fas and TNF can generate ceramide and activate proteases (Sect. 6.6), and Bax can direct cell death without protease activation [181]. In addition Bcl-2 family members and antioxidants can provide protection against TNF mediated cell death [106]. While recent progress has been made in the identification of molecular mediators, it is evident that significant gaps in our knowledge remain

6.1
The *bcl-2* Gene Family

The *bcl-2* gene family includes both positive (*bax, bcl-xs, bad, bik*) and negative regulators of apoptosis (*bcl-2, bcl-xL, mcl-1*). Family members interact to form homodimers and heterodimers with other related or non-related proteins. Cell susceptibility to apoptosis is thought to be largely influenced by the relative ratios of pro-apoptotic and anti-apoptotic *bcl-2* family members. The prototype of the family, *bcl-2*, was originally cloned from the t(14:18) translocation breakpoint associated with follicular lymphoma and was subsequently identified as a negative regulator of apoptosis analogous to its homologue, *ced-9* in *C. elegans*. Over-expression of *bcl-2* has been shown to protect a variety cell lines from apoptosis induced by a broad range of signals, including growth factor deprivation, UV irradiation, oxidative stress, cytotoxic drugs, and heat shock ([111] and Chap. 4 of this volume). The ability of Bcl-2 to protect against such a broad range of inducing signals suggests that the biochemical functions of this family form a late control point prior to the execution stage of apoptosis. Bcl-2 does not protect against all death inducing signals, for example apoptosis induced by cytotoxic T-lymphocytes [112], suggesting the presence of alternative pathways which circumvent this control point. The biochemical mechanisms by which *bcl-2* family members influence apoptosis are presently unknown, although the location of *bcl-2* (mitochondrial membrane, endoplasmic reticulum and nuclear envelope), functional dependence upon membrane anchoring [113], and the structural homology of family members to the diphtheria toxin channel suggests that they may form ion channels *in vivo* [114].

6.2
p53

The *p53* tumour suppressor gene encodes a protein which is critical for maintaining the integrity of the genome. This is evidenced by the fact that families harbouring a *p53* mutation have a high probability (50%) of developing cancer at an early age, and *p53* deficient mice are highly susceptible to spontaneous tumour formation. In addition, *p53* mutations occur in more than half of all human tumours, and represent the most common genetic alteration in human cancer.

In response to sub-lethal concentrations of various types of genotoxic insults (e.g. UV light, gamma irradiation, DNA damaging drugs), p53 directs cell cycle arrest in G1, allowing DNA repair to take place prior to further replication. In the event of excessive DNA damage, p53 mediated apoptosis follows cell cycle arrest [79, 80]. p53 has also been implicated in mediating the physiological response of cells to survival factors [115]. Endogenous p53 expression was shown to significantly accelerate apoptosis in myeloid 32D cells upon withdrawal of IL-3 [116], and anti-sense p53 oligonucleotides reduced the level of apoptosis in factor dependent leukaemia cells after withdrawal of growth factor [117].

The mechanisms involved in p53 induced cell cycle arrest are much better understood than those involved in the induction of apoptosis. The two functions of p53 may be uncoupled suggesting independent regulation. For example apoptosis can proceed without G1 arrest in p21 deficient mice [118, 119] and p53 mediated apoptosis can be blocked by Bcl-2 without affecting cell cycle arrest [120]. The latter study also positions p53 upstream of the Bcl-2 control point (see Fig. 3). p53 mediated cell cycle arrest is dependent upon transcription, whereas apoptosis can be induced by both transcriptionally dependent [121, 122] and independent means [123, 124]. Activation of *bax* and *Fas* genes and/or repression of the *bcl-2* gene may play an important role in transcription dependent p53 mediated apoptosis [125–127].

p53 is not required for glucocorticoid-mediated apoptosis of thymocytes [128] and cell lines which do not express p53 are still capable of undergoing apoptosis. Apoptosis can therefore proceed both p53 dependent and independent pathways. p53 is not required for apoptosis during normal development, suggesting that its primary role may be as a sensor of DNA damage [14].

6.3
myc

The *myc* family of proto-oncogenes encode DNA binding proteins which can heterodimerise to form transcriptional activators or repressors. They have been implicated in the regulation of genes involved in proliferation, differentiation and more recently apoptosis [129].

Normal proliferating fibroblasts maintain c-Myc expression throughout the cell cycle. When the cells are deprived of serum or growth factors c-Myc is rapidly downregulated, leading to cell cycle arrest in G_1. Constitutive high level expression of c-Myc abolishes cell cycle arrest in fibroblasts following serum deprivation, and results in concomitant proliferation and apoptosis [130]. Although Myc dependent apoptosis requires heterodimerisation with its partner Max, apoptosis can proceed in the absence of protein synthesis. This suggested that Myc must be continuously implementing the molecular machinery required for apoptosis, but that its instigation relies upon other factors. It was subsequently shown that the insulin-like growth factors and PDGF could inhibit Myc induced apoptosis in low serum [131]. In addition, anti-apoptotic proteins such as Bcl-2, v-Abl and mutant P53 can inhibit Myc induced cell death, enabling proliferation to be the primary biological consequence of Myc de-regulation [132–134]. *In vivo* co-operation between *myc* and *bcl-2* has been demonstrated in experiments with transgenic mice. Mice transgenic for *myc* alone did not show hyperplasia despite an augmentation in proliferation, whereas *myc/bcl-2* double transgenics rapidly developed malignant lymphomas [135]. Thus the ability of Myc to induce two separate and functionally conflicting pathways necessitates the presence of a survival cytokine or gene, and provides a rationale for oncogene co-operation in cellular transformation [136].

6.4
ras Genes

Proteins encoded by *ras* gene family members are GTP-binding, signal transducing proteins which have also been associated with both mitogenic and apoptotic responses. *ras* genes become oncogenic when activated by mutation, and can cooperate with other genes such as *p53, myc, E1A*, and *SV40* in cellular transformation [137]. In the presence of IRF-1 (interferon regulatory factor 1) cDNA, cells with activated *ras* undergo apoptosis, suggesting that IRF-1 may act like a tumour suppressor gene as it can mediate the death of an oncogene activated cell [138]. Other studies have suggested that the consequence of *ras* activation is dependent upon the activity of protein kinase C (PKC). Activated *ras* induced apoptosis in a T-lymphoblastoid cell line, when PKC activity was supressed. Apoptosis induced by Ras in this system could be blocked by Bcl-2 [139]. In contrast, other groups have reported that activation or over-expression of H-*ras* could have a protective effect on apoptosis [140]. The effects of H-*ras* may therefore be largely dependent on the presence of other regulatory proteins. R-Ras is a Ras related, Bcl-2 interacting protein [141], which can promote apoptosis in growth factor deprived cells by a mechanism which is suppressible by Bcl-2 [142].

6.5
Abl Tyrosine Kinases

The oncogenic effects of Abl tyrosine kinases have been reported to be a consequence of their inhibitory effect on apoptosis [143]. Both v-Abl and Bcr-Abl can rescue growth factor dependent cell lines from apoptosis induced by factor withdrawal [144, 145] and protect myeloid cells from apoptosis induced by a variety of cytotoxic drugs and Fas ligation [90, 146]. Bcr-Abl has been reported to prevent apoptosis in factor dependent murine haematopoietic cells by inducing a Bcl-2 expression pathway [147], although another study has suggested that the inhibition of apoptosis by v-Abl is independent of both Bcl-2 and Bax expression [148]. Activated Ras, possibly mediated by the direct interaction of Abl and Ras regulatory proteins, is required for the oncogenic activity of Abl [143].

6.6
The Fas and TNF Receptor Family

Many cellular responses such as proliferation, differentiation and survival are mediated by extra-cellular signals (cytokines, adhesion molecules etc.) acting through specific cell surface receptors. It is perhaps unsurprising therefore that specific receptor-ligand binding can also mediate cell death. Fas ligand (FasL) and tumour necrosis factors (TNFs) can rapidly induce apoptosis in target cells expressing their receptors. Fas and TNF receptors belong to the expanding tumour necrosis factor receptor (TNFR) family of surface receptors. The receptors are all transmembrane proteins which possess characteristic cysteine-rich repeats in their extracellular domains. Most of the family members, including nerve growth factor receptor (NGF), CD40, CD27 and CD30 [149], transduce proliferation or differentiation stimulatory signals, although two recent additions DR3/Wsl-1 [150] and CAR-1 [151] are also capable of inducing cell death.

Both the Fas and TNF receptors are functionally dependent upon a specific sequence of 65 amino acids in their carboxyl terminus referred to as the "death domain" (DD). The death domain recruits other DD containing proteins (FADD/MORT1-FasR and TRADD-FADD/MORT1-TNFR) to the receptor complex. Signalling from these complexes is thought to be mediated by another DD containing protein termed FLICE (FADD like ICE), which possesses an additional functional domain with homology to the ICE like proteases (see below). FLICE is therefore classed as a member of the caspase family and is referred to as caspase 8 [152]. Caspase 8 is thought to provide a crucial link between Fas and

TNF receptor complexes, and downstream effector ICE-like proteases (caspases) (reviewed in Chaps. 3 and 5 in this volume). Other biochemical events which have been implicated in Fas signalling include activation of phosphatidylcholine-specific phospholipase C (PC-PLC), and production of ceramides by an acidic sphingomyelinase [93]. The mechanism by which ceramides may activate downstream proteases in presently unknown (see also Sect. 5.4.3).

FasR is widely expressed on normal and malignant cell types, and is abundantly expressed in the thymus, liver, heart and kidney. The consequences of Fas deficiency are, however, primarily manifested in the immune system, where Fas receptor ligation is thought to be important for the clonal deletion of auto-reactive lymphocytes, and in the down-regulation of an immune response. Children carrying a heterozygous defect in the FasR, and mouse models harbouring mutations in either FasL or FasR, display lymphadenopathy and auto-immune diseases (see Sect. 8.1 on cell death and disease).

FasL is upregulated on cytotoxic T-cells (T_c-cells) in response to T-cell receptor interaction with viral antigens. The interaction between FasL on the T_c-cell and Fas R on the infected cell results in target cell apoptosis [39, 149]. T_c-cells can also kill target cells by a second mechanism resulting in the formation of membrane pores and release of granzyme B, which can activate caspases directly [25].

6.7
Caspase/Protease Family

The cloning of the cell death promoter CED-3 in *C. elegans*, revealed significant homology to interleukin-1β-converting enzyme (ICE), and provided the first indication that proteases may play a central role in apoptosis [12]. Several mammalian homologues were subsequently identified including Nedd2/Ich-1, CPP32, $ICE_{rel\ II}$/TX/Ich-2, $ICE_{rel\ III}$, Mch2 ([153] and Chap. 5 in this volume). The identification of some of these proteases by a number of independent groups resulted in multiple names for the same protein and confusing terminology. Consequently ICE like proteases have now been designated as caspases 1 – 10 [152].

Caspases are produced as inactive pro-enzymes (zymogens). They are activated by proteolytic cleavage, which generates two subunits that form a tetramer. Active caspases cleave their substrates after an aspartate residue in specific recognition sequences. Recognition sequences have been identified in several substrates including U1 – 70 kDa small ribonucleoprotein, lamin B1, poly-(ADP) ribose polymerase, as well as in caspases themselves, suggesting sequential activation in a proteolytic cascade.

The genetic evidence from *C. elegans* implying a central effector role for proteases was supported by the fact that over-expression of caspases induces apoptosis in a variety of cell types. Moreover, specific inhibitors of caspases block apoptosis induced by a wide range of stimuli. Caspase inhibitors include viral gene products and short synthetic polypeptides (see above-mentioned reviews and references therein). CrmA is a caspase inhibitor encoded by cowpox virus which preferentially inhibits caspase 1 (ICE) [154] and *p35* is a bacilovirus gene which inhibits various members of the caspase family [155]. CrmA and p35 can inhibit apoptosis induced by both Fas and TNF suggesting that caspases (specifically caspase 1) are downstream mediators of their cytotoxicity [156]. An analysis of protease activation following Fas ligation found that caspase 1 was transiently activated prior to caspase 3, and that activation of caspase 3 was dependent upon the activity of caspase 1, thus supporting the notion of a sequentially activating cascade.

Several studies have shown that inhibition of apoptosis by over-expression of Bcl-2 or Bcl-xL also inhibits activation of caspases. It has also been demonstrated that Bcl-2 cannot prevent substrate cleavage by active caspases, thus placing Bcl-2 proximal to caspases in the molecular pathway to cell death [157].

6.8
CED-4: The Missing Link?

Although it has been established that the *bcl-2* and caspase families comprise key regulatory proteins in the control and execution phases of apoptosis, the molecular link between these two families remains elusive. They both represent mammalian equivalents to key regulatory proteins in *C. elegans*, CED-9 and CED-3 respectively. However a mammalian homologue for CED-4 (also essential for cell death in *C. elegans)* has never been identified. Recent evidence has demonstrated that CED-4 can provide a direct molecular link between Bcl-2 and caspases in mammalian cells. Over-expression of CED-4 in mammalian cells induces apoptosis which can be blocked by Bcl-x_L and by caspase inhibitors, suggesting that Bcl-2 family members mediate CED-4 activation of caspases. Direct physical interaction was also demonstrated between Ced-4 and Ced-9 (or Bcl x_L), and between Ced-4 and either Ced-3, FLICE or ICE, but not CPP32 or Mch-2 which have smaller pro-domains. The same study also suggested that CED-4 could simultaneously interact with both CED-9 and CED-3 [158]. Immunostaining techniques indicated that CED-4 was predominately localised in the cytoplasm, but that co-expression with CED-9 resulted in its re-distribution from the cytosol to intra-cellular membrane sites. It was proposed that the anchoring of CED-4 to CED-9 (Bcl-2) may restrain its ability to activate CED-3 [159].

It is likely that, as in the case of CED-9 and CED-3, equivalent mammalian mechanisms employ a host of analogous regulatory proteins in apoptosis pathways. Release of cytochrome *c* [160] and AIF (apoptosis inducing factor) [161] have also been suggested to provide molecular links between the Bcl-2 and caspase families.

7
Viral Exploitation of Cell Death Mechanisms

While scientists have only recently appreciated the biological significance of apoptosis, viruses have been selfishly manipulating its molecular mechanisms for thousands of years. Evidence from studies with de-regulated oncogenes (e.g. *myc* or *ras*) suggests that the corruption of a cell's proliferative machinery may activate a default program for cell death, and that this may be a protective mechanism against viral propagation or tumorogenesis. It is not surprising therefore that viruses have acquired anti-apoptotic genes (probably from an ancestral host) in order to circumvent the altruistic suicide of its host.

For example the DNA tumour virus, adenovirus, produces an oncoprotein E1 A which can activate the transcriptional machinery for proliferation, but in doing so, also activates *p53* dependent apoptosis [162]. Consequently, adenovirus has procured two additional proteins which inhibit apoptosis through interaction with P53 and Bax [163]. Human papilloma virus (HPV) encodes a protein product, E7, which also promotes cellular proliferation and an additional HPV protein, E6, which inhibits apoptosis by promoting the intracellular degradation of *p53*.

Other anti-apoptotic genes encoded by viruses include the Epstein Barr virus BHRF1 protein (a Bcl-2 homologue) [164], and the aforementioned caspase inhibitors. CrmA protein from cowpox virus inactivates the ICE protease [165] whereas p35 and IAP proteins from baciloviruses inhibit a broader range of caspases [166]. Interestingly, a human homologue of IAP has recently been described. Neuronal Apoptosis Inhibiting Protein (NAIP) is one of a pair of genes which are mutated in Type 1 spinal muscular atrophy. As this disease is characterised by motor neuron degeneration, it raises the possibility that NAIP is an important anti-apoptotic gene in neurons [167].

Clearly viruses have identified the most significant molecular control points in apoptosis pathways and have effectively exploited them for mercenary reasons. This has necessitated the parallel evolution of host defence mechanisms which can initiate apoptosis in infected cells by a mechanism which circumvents many of the upper control points in the apoptotic pathway. Thus mammalian cytotoxic T-lymphocytes initiate apoptosis in infected or malignant cells both by receptor ligation directly activating FLICE and ICE (Fas pathway), and direct activation of downstream proteases (e. g. CPP32) by granzyme B [25].

8
Implications for Biomedical and Biopharmaceutical Research

The knowledge that cell death is mediated by a highly conserved, genetically defined series of biochemical events presents the possibility of its manipulation and provides an impetus for more intensive research. An ability to modulate the capacity of a cell to initiate a program for death would be an invaluable asset in the development of new therapeutic regimes, or in the optimisation of cell culture productivity in industry.

8.1
Cell Death and Disease

The importance of apoptosis in both specialised functions and in normal tissue homeostasis is evident from the disastrous consequences of its deregulation. The deregulation of apoptosis is now known to play a major role in the pathogenesis of a variety of diseases. Excessive apoptosis has been associated with the progression of various neurodegenerative disorders and acquired immunodeficiency syndrome (AIDS), whereas insufficient apoptosis plays a major role in the pathogenesis of auto-immune disease, inflammatory conditions, and the development of malignancy. [38, 136, 168].

Extensive apoptosis features prominently in the pathology of several neurodegenerative conditions, for example Alzheimer's, Parkinson's, Huntington's amyotrophic lateral sclerosis and retinitis pigmentosa [169–171]. Selective elimination of subsets of neuronal cells by apoptosis results in inappropriate cell depletion and disease progression. Excessive apoptosis has also been proposed to account for the pathological effects of HIV infection, leading to T-cell depletion and consequent inability to mount an immune response [172]. Hepatitis B and C viruses also result in the death of vital cells, due to the induction of apoptosis in the liver by subsets of cytotoxic T-cells. Liver damage in mice could be prevented by Fas neutralising molecules or by over-expression of Bcl-2 [173].

Both Fas and FasL ligand were first identified by researchers investigating the genetic abnormality in auto-immune diseases in mice [174]. It is now apparent that an inherited defect in Fas can lead to a similar auto-immune disorder in humans [175]. The production of auto-antibodies, lymph node enlargement and expanded populations of CD4⁻CD8⁻ T-cells are thought to be a consequence of an inability to induce apoptosis in peripheral lymphocytes [174].

The first anti-apoptotic gene to be identified in humans was the *bcl-2* gene. The gene is overexpressed as a result of the t(14:18) translocation found in most follicular lymphomas and some diffuse large cell lymphomas. Experiments with transgenic mice confirmed the role of Bcl-2 in the pathogenesis of the disease and also highlighted the importance of gene cooperation in transformation (see Sects. 6.1 and 6.3). DNA tumour viruses have also demonstrated the requirement for both proliferative and anti-apoptotic lesions in the acquisition of malignancy (see Sect. 7). In chronic myeloid leukaemia (CML) a chromosome translocation results in the production of a chimeric protein Bcr-Abl, which was subsequently shown in *in vitro* experiments to be a potent inhibitor of apoptosis (Sect. 6.5). The attainment of an "anti-apoptotic state" in malignancy not only contributes to the progression of disease but is also likely to play a major role in the development of resistance to chemotherapeutic drugs [38].

It is possible that replenishment of survival factors, or inhibition of death inducing receptor signals, may provide future novel targets for therapy in degenerative diseases. Conversely, depletion of survival signals, or inactivation of over-expressed or abnormal anti-apoptotic genes by antisense therapy, may be future options in the treatment of malignancy. An interesting feature of both HIV induced apoptosis, and apoptosis in neurodegenerative disease is the increase in oxidative stress during these conditions [176]. An imbalance in cellular redox status has also been reported to promote tumour development [177]. It is anticipated that a better understanding of both the genetic and biochemical mediators in apoptosis pathways should facilitate the development of new therapeutic options for the treatment of the various diseases in which de-regulated apoptosis plays a significant role (see also Chap. 8 in this volume).

8.2
Cell Death in Large Scale Cultures

A variety of animal cell lines are employed in the large scale production of commercially important biological molecules such as antibodies or vaccines. Optimisation of cell culture conditions in industrial bioreactors is important for both quantitative and qualitative purposes. Cell death due to the adverse conditions prevailing in bioreactors is often a significant limiting factor in culture productivity. Recent research has shown that death induced under a range of stressful conditions such as mechanical agitation, nutrient depletion and cytotoxin accumulation is initially predominately apoptotic [178, 179]. Apoptotic cells in non-physiological environments subsequently lose membrane integrity and undergo secondary necrosis. Both primary and secondary necrosis are undesirable as not only do they reduce the number of productive cells, they also result in the release of proteases which can damage the cellular derived product.

Strategies for productivity optimisation have previously focused on cell growth, largely because it was considered that death was passive and thus unavoidable. The recent influx in information regarding cell death regulation has however presented new opportunities for the enhancement of cell survival. It may be possible to promote cell viability by improving culture media with specific survival factors, or anti-oxidant molecules. Alternatively, the generation of more robust cell lines by genetic manipulation may prove to be a more cost effective long term option. It has already been established that the expression of Bcl-2 in a variety of non-commercial cell lines can protect them from a wide range of cytotoxic insults. In addition, a recent study has demonstrated that the stable expression of Bcl-2 in a commercially important hybridoma cell line produced a more adaptable cell line and significantly increased productivity ([180] and Chap. 9 in this volume).

It is possible that other negative regulators of apoptosis for example Bcl-xL, Mcl-1, or viral inhibitors such as p35 may also prove to be powerful tools for the manipulation of apoptotic induction thresholds in large cell cultures.

9
Conclusions and Future Perspectives

For many years the misconception of cell death as a purely passive process was accompanied by a passive attitude with regard to its biological significance. Recent research has however dispelled such attitudes and generated a surge of interest in the subject. In particular, expectations have been raised for novel therapeutic options and productivity improvements in commercially important cellular derived proteins.

Appropriate exploitation of cell death mechanisms will undoubtedly require a more intricate knowledge of the various control points in apoptosis induction and execution. For example, what are the immediate sensors of damage or death inducing signals, and how do these sensors relay information to the initiators of apoptotic pathways? It is possible that a better knowledge of specific pathways may allow the development of more selective and thus therapeutically useful inhibitors or inducers of apoptotic cell death. Alternatively, lessons from viruses suggest that the further downstream the target, the greater the scope and efficiency of the inhibitor. Such an approach may be more beneficial in bio-reactors where a variety of factors influence cell death and only one cell type has to be considered.

It is notable that many of the diseases in which deregulation of apoptosis is thought to play a significant role are both fatal and incurable. The rewards for apoptosis research are undoubtedly high and justify the explosion in interest and expenditure in the field.

Acknowledgements. We wish to thank the EU Biotech programme, Heath Research Board of Ireland and the Irish Cancer Society for their generous financial support.

Notes added in proof: A protein called DFF (DNA fragmentation factor) has been purified from human cells, cloned and sequenced. It is activated by caspase 3 and results in DNA fragmentation. Liu X and Zou H (1997) Cell 89:175–184. A caspase activated DNase – CAD and its inhibitor ICAD have been isolated from mouse cells, cloned and sequenced. ICAD and DFF are highly homologous. Enari M, Sakahira H, Yokoyama H, Okawa K, Iwamatsu A, Nagata S (1988) Nature 391:43–50.

A human homologue of CED-4 has now been identified (Zou et al., 1997 Cell vol. 90:405–413). The protein is called Apaf-1 (apoptotic protease activating factor-1), and has regions of homology to both CED-4 and CED-3. Binding of catochrome C to Apaf-1 is required to trigger the activation of caspase-3.

10
References

1. Clarke PGH, Clark S (1996) Anat Embryol 193:81
2. Lockshin RA, Williams CM (1964) J Insect Physiol 10:643
3. Saunders JW (1966) Science 154:604
4. Tata JR (1966) Dev Biol 13:77
5. Munk A (1971) Perspect Biol Med 14:265
6. Lockshin RA (1969) J Insect Physiol 15:1505
7. Kerr JFR, Wyllie AH, Currie AH (1972) Br J Cancer 26:239
8. Wyllie AH, Kerr JFR, Currie AR (1980) 68:251
9. Ellis HM, Horvitz HR (1986) Cell 44:817
10. Ellis RE, Yuan JY, Horvitz HR (1991) Ann Rev Cell Biol 7:663
11. Williams GT, Smith CA, McCarthy NJ, Grimes EA (1992) Trends Cell Biol 2:263
12. Yuan J, Shaham S, Ledoux S, Ellis HM, Horvitz HR (1993) Cell 75:641
13. Hengartner MO, Horvitz HR (1994) Cell 76:665
14. White E (1996) Genes and Dev 10:1
15. Chinnaiyan A, Dixit V (1996) Curr Biol 6:555
16. Cohen JJ, Duke RC, Fadok VA, Sellins KS (1992) Ann Rev Immunol 10:267
17. Schwarz LM, Smith SW, Jones MEE, Osborne BA (1993) 90:980
18. Jacobson MD, Weil M, Raff MC (1997) Cell 88:347
19. Vaux DL, Weissman IL, Kim SK (1992) Science 258:1955
20. Cornillon S, Foa C, Davoust J, Buonavista N, Gross JD, Goldstein P (1994) J Cell Sci 107:2691
21. Ameisen JC, Idziorek T, Billaut-Mulot O, Loyens M, Tissier JP, Potentier A (1995) Cell Death and Diff 2:285
22. Haecker G, Vaux DL (1994) Trends Biochem Sci 19:99
23. Yu YT, Snyder L (1994) Proc Natl Acad Sci USA 91:802

24. Snyder L (1995) Mol Microbiol 15:415
25. Greenberg AH (1996) Cell Death and Diff 3:269
26. Arends MJ, Morris RJ, Wyllie AH (1990) Am J Pathol 136:593.
27. Kerr JFR, Searle J, Harmon BV, Bishop CJ (1987) Apoptosis. In: Potten CS (ed) Perspectives on mammalian cell death. Oxford University Press, Oxford, New York, Tokyo. p 93
28. Coucouvanis E, Martin GR (1995) Cell 83:279
29. Glucksmann A (1951) Biol Rev 26:59
30. Lewinson D, Silbermann M (1992) Anat Rec 233:504
31. Furtwangler JA, Hall SH, Koskinen ML (1985) Acta Anat 124:74
32. Penfold PL, Provis JM (1986) Graefe's Arch Clin Exp Ophthalmol 224:549
33. Barde YA (1989) Neuron 2:1525
34. Oppenheim RW (1991) Annu Rev Neurosci 14:453
35. White E, Livanos EM, Tlsty TD (1994) Genes Dev 8:666
36. Kuida K, Zheng TS, Na S, Kuan C-Y, Yang D, Karasuyama H, Rakic P, Flavell RA (1996) Nature 384:368
37. Raff MC (1992) Nature 356:397
38. McKenna SL, Cotter TG (1997) Adv Cancer Res 71:121
39. Berke G (1995) Cell 81:9
40. Cohen JJ, Duke RC (1984) J Immunol 132:38
41. Oberhammer F, Wilson JW, Dive C, Morris ID, Hickman JA, Wakeling AE, Walker PR, Sikorska M (1993) EMBO J 12:3670
42. Brown DG, Sun XM, Cohen GM (1993) J Biol Chem 268:3037
43. Walker PR, Pandey S, Sikorska M (1995) Cell Death Diff 2:97
44. Filipski J, Leblanc J, Youdale T, Sikorska M, Walker PR (1990) EMBO J 9:1319
45. Walker PR, Sikorska M (1986) Biochemistry 25:3839
46. Peitsch MC, Polzar B, Stephen H, Crompton T, Macdonald HR, Mannherz HG, Tschopp J (1993) EMBO J 12:371
47. Barry MA, Eastman A (1993) Arc Biochem Biophys 300:440
48. Gaido ML, Cidlowski JA (1991) J Biol Chem 266:18,580
49. Deng G, Podack ER (1995) FASEB J 9:665
50. Tanuma S, Shiokawa D (1994) Biochem Biophys Res Comm 203:789
51. Montague JW, Gaido ML, Frye C, Cidlowski JA (1994) J Biol Chem 269:18,877
52. Nikonova LV, Beletsky IP, Umansky SR (1993) Eur J Biochem 215:893
53. Walker PR, Sikorska M (1994) Biochem Cell Biol 72:615
54. Hughes FM, Cidlowski JA (1994) Cell Death Diff 1:11.
55. Anzai N, Kawabata H, Hirama T, Masutani H, Ueda Y, Yoshida Y, Okuma M (1995) Blood 86:917
56. Martin SJ, Green DR (1995) Cell 82:349
57. Zhivotovsky B, Burgess DH, Orrenius S (1996) Experientia 52:968
58. Jacobsen MD, Burnett JF, Raff MC (1994) EMBO J 13:1899
59. Martin SJ, Newmeyer DD, Mathias S, Farschon DM, Wang HG, Reed JC, Kolesnick RN, Green DR (1995) EMBO J 14:5191
60. Kayalar C, Ord T, Testa MP, Zhong LT, Bredesen DE (1996) Proc Nat Acad Sci USA 93:2234
61. Martin SJ, O'Brien GA, Nishioka WK, McGahon AJ, Mahboubi A, Saido TC, Green DR (1995) J Biol Chem 270:6425
62. Hockenberry DM, Oltvai ZN, Yin XM, Milliman CL, Korsmeyer SJ (1993) Cell 75:241
63. Zamzami N, Susin SA, Marchetti P, Hirsch T, Gomez-Monterray I, Castedo M, Kroemer G (1996) J Exp Med 183:1533
64. Jacobson MD, Burnett JF, King MD, Miyashita T, Reed JC, Raff MC (1993) Nature 361:365
65. Jacobson MD, Raff MC (1995) Nature 374:814
66. Fesus L, Thomazy V, Falus A (1987) FEBS Lett 224:104
67. Savill JS, Hogg N, Ren Y, Haslett C (1992) J Clin Invest 90:1513
68. Martin SJ, Reutlingsperger CP, McGahon AJ, Rader AJ, van-Schie RC, LaFace DM, Green DR (1995) J Exp Med 182:1545
69. Yonehara S, Ishiti A, Yonehara M (1989) J Exp Med 169:1747

70. Itoh N, Yonehara S, Ishii A, Yonehar M, Mizushima S, Sameshima M, Hase A, Seto Y, Nagata S (1991) Cell 66:233
71. Collins MKL, Lopez Rivas A (1993) TIBS 18:307
72. Ruoslahti E, Reed JC (1994) Cell 77:477
73. Kaufmann SH (1989) Cancer Res 49:5870
74. Hannun YA (1997) Blood 89:1845
75. McKenna SL, Padua RA (1997) Br J Haematol 96:659
76. Calsou P, Salles B (1993) Cancer Chemother Pharmacol 32:85
77. Fox M, Roberts JJ (1987) Canc and Metas Rev 6:261
78. Feige U, Polla B (1994) Experientia 50:979
79. Morgan SE, Kastan MB (1997) Adv. Cancer Res 71:1
80. Yonish-Rouach E (1996) Experientia 52:1001
81. de Murcia G, de Murcia JM (1994) TIBS 19:172
82. Lindahl T, Satoh MS, Poirier GG, Klungland A (1995) TIBS 20:405
83. Satoh MS, Lindahl T (1992) Nature 356:356
84. Hale AJ, Smith CA, Sutherland LC, Stoneman VEA, Longthorne VL, Culhane AC, Williams GT (1996) Eur J Biochem 236:1
85. McConkey DJ, Nicoterra P, Orrenius S (1994) Immunol Rev 142:343
86. Buttke TM, Sandstrom PA (1994) Immunol Today 15:7
87. Slater AFG, Stefan C, Nobel I, van del Dobbelsteen DJ, Orrenius S (1996) Cell Death Diff 3:57
88. Erpel T, Courtneidge SA (1995) Curr Opin Cell Biol 7:176
89. Uckun FM, Tuel-Ahlgren L, Song CW, Waddick K, Myers DE, Kirihara J, Ledbetter JA, Schieven GL (1992) Proc Natl Acad Sci USA 89:9005
90. McGahon A, Bissonnette R, Schmidt M, Cotter KM, Green DR, Cotter TG (1994) Blood 83:1179
91. Shi L, Nishioka WK, Th'ng J, Bradbury EM, Litchfield DW, Greenberg AH (1994) Science 263:1143
92. Obeid LM, Linardic CM, Karolak LA, Hannun YA (1993) Science 259:1769
93. Quintans J, Gottschalk AR (1994) The Immunologist 2:185
94. Hannun YA, Obeid LM (1995) TIBS 20:73
95. Lavin MF, Watters D, Qizhong S (1996) Experientia 52:979
96. Sawai H, Okazaki T, Yamamoto H, Okano H, Takeda Y, Tashima M, Sawada H, Okuma M, Ishikura H, Umechara H, Domae N (1995) J Biol Chem 270:27,326
97. Bessho R, Matsubara K, Kubota M, Kuwakado K, Hirota H, Wakazono Y, Wei Lin Y, Okuda A, Kawai M, Nishikomori R, Heike T (1994) Biochem Pharm 48:1833
98. McGowan AJ, Ruiz-Ruiz MC, Gorman AM, Lopez Rivas A, Cotter TG (1996) FEBS Lett 392:299
99. Lennon SV, Martin SJ, Cotter TG (1991) Cell Prolif 24:203
100. Torres-Roca JF, Lecoeur H, Amatore C, Gougeon ML (1995) Cell Death Diff 2:309
101. Nobel CSI, Kimland M, Lind B, Orrenius S, Slater AFG (1995) J Biol Chem 270:26,202
102. Behl C, Davis JB, Lesley R, Schubert D (1994) Cell 77:817
103. Schulze-Osthoff K, Bakker AC, Vanhaesebroeck B, Beyaert R, Jacob WA, Fiers W (1992) J Biol Chem 267:5317
104. Slater AFG, Stefan C, Nobel I, Maellaro E, Bustamante J, Kimland M, Orrenius S (1995) Biochem J 306:771
105. Verhaegen S, McGowan AJ, Brophy AR, Fernandes RS, Cotter TG (1995) Biochem Pharm 50:1021
106. Briehl MM, Baker AF (1996) Cell Death Diff 3:63
107. Jacobson MD, Raff MC (1995) Nature 374:814
108. Shimizu S, Eguchl Y, Kosaka H, Kamiike W, Masuda H, Tsujimoto Y (1995) Nature 374:811
109. Richter C (1993) FEBS Lett 325:104
110. Grune T, Reinheckel T, Joshi M, Davies KJA (1995) J Biol Chem 270:2344
111. Yang E, Korsmeyer SJ (1996) Blood 88:386
112. Vaux DL, Aguila HL, Weissman IL (1992) Int Immunol 4:821

113. Nguyen M, Branton PE, Walton PA, Oltvai ZN, Korsmeyer SJ, Shore GC (1994) J Biol Chem 269:16,521
114. Muchmore SW, Sattler M, Liang H, Meadows RP, Harlan JE, Yoon HS, Nettesheim D, Chang BS, Thompson CB, Wong SL, Ng S-C, Fesik SW (1996) Nature 381:335
115. Lin Y, Benchimol S (1995) Mol Cell Biol 15:6045
116. Blandino G, Scardigli R, Rizzo MG, Crescenzi M, Soddu S, Sacchi A (1995) Oncogene 10:731
117. Zhu Y-M, Bradbury DA, Russel NH (1994) Br J Cancer 69:468
118. Deng C, Zhang P, Harper JW, Elledge SJ, Leder P (1995) Cell 82:675
119. Brugarolas J, Chandrasekaran C, Gordon JI, Beach D, Jacks T, Hannon JG (1995) Nature 377:552
120. Guillouf C, Grana X, Selvakumaran M, De Luca XA, Giordano A, Hoffman B, Liebermann DA (1995) Blood 85:2691
121. Yonish-Rouach E, Deguin V, Zaitchouk T, Breugnot C, Mishal Z, Jenkins JR, May E (1995) Oncogene 11:2197
122. Sabbatini P, Lin J, Levine AJ, White E (1995) Genes and Dev 9:2184
123. Caelles C, Helmberg A, Karin M (1994) Nature 370:220
124. Haupt Y, Rowan S, Shaulian E, Vousden KH, Oren M (1995) Genes and Dev 9:2170
125. Miyashita T, Reed JC (1995) Cell 80:293
126. Owen-Schaub LB, Zhang W, Cusack JC, Angelo LS, Santee SM, Fujiwara T, Roth JA, Deisseroth AB, Zhang W-W, Kruzel E (1995) Mol Cell Biol 15:3032
127. Miyashita T, Harigai M, Hanada M, Reed JC (1994) Cancer Res 54:3131
128. Clark AR, Purdie CA, Harrison DJ, Morris RG, Bird CC, Hooper ML, Wyllie AH (1993) Nature 362:849
129. Amati B, Land H (1994) Curr Opin Genet Dev 4:102
130. Evan GI, Wyllie AH, Gilbert CS, Littlewood TD, Land H, Brooks M, Waters CM, Penn LZ, Hancock DC (1992) Cell 69:119
131. Harrington EA, Bennett MR, Fanidi A, Evan GI (1994) EMBO J 13:3286
132. Bissonnette RP, Echeverri F, Mahboubi A, Green DR (1992) Nature 359:552
133. Fanidi A, Harrington EA, Evan GI (1992) Nature 359:554
134. Lotem J, Sachs L (1995) Proc Natl Acad Sci USA 92:9672
135. Cory S, Harris AW, Strasser A (1994) Philos-Trans-R-Soc-Lond-B-Biol-Sci 345:289
136. Green DR, Bissonnette RP, Cotter TG (1994) Apoptosis and Cancer. In: De Vita VT, Hellman S, Rosenberg SA (eds) Important advances in oncology. Lippincott Company, Philadelphia, p 37
137. Barbacid M (1987) Ann Rev Biochem 56:779
138. Tanaka N, Ishihara M, Kitagawa M, Harada H, Kimura T, Matsuyama T, Lamphier MS, Aizawa S, Mak TW, Taniguchi T (1994) Cell 77:829
139. Chen CY, Faller DV (1995) Oncogene 11:1487
140. Lin HJL, Eivner V, Prendergast GC, White E (1995) Mol Cell Biol 15:4536
141. Fernandez-Sarabia MJ, Bischoff JR (1994) Nature 366:274
142. Wang H-G, Millan JA, Cox AD, Der CJ, Rapp UR, Beck T, Zha H, Reed JC (1995) J Cell Biol 129:1103
143. Fernandes RS, Gorman AM, McGahon A, Lawlor M, McCann S, Cotter TG (1996) Leukemia 10:S17
144. Rovera G, Valtieri M, Maullo F, Reedy EP (1987) Oncogene 1:29
145. Daley GQ, Baltimore D (1988) Proc Natl Acad Sci USA 85:9312
146. McGahon AJ, Nishioka WK, Martin SJ, Mahboubi A, Cotter TG, Green DR (1995) J Biol Chem 270:22,625
147. Sanchez-Garcia I, Grutz G (1995) Proc Natl Acad Sci USA 92:5287
148. McGahon AJ, Martin SJ, Jin Yoo N, Bissonette, RP, Harigai M, Cotter TG, Reed JC, Green DR (1997) Submitted
149. Nagata S, Goldstein P (1995) Science 267:1449
150. Chinnaiyan AM, O'Rourke K, Yu G-L, Lyons RH, Garg M, Duan RD, Xing L, Gentz R, Ni J, Dixit VM (1996) Science 274:990

151. Brojatsch J, Naughton J, Rolls MM, Zingler K, Young JAT (1996) Cell 87:845
152. Alnemri ES, Livingston DJ, Nicholson DW, Salvesen G, Thornberry NA, Wong WW, Yuan J (1996) Cell 87:171
153. Kumar S, Lavin MF (1996) Cell Death Diff 3:255
154. Miura M, Zhu H, Rotello R, Hartwieg EA, Yuan J (1993) Cell 75:653
155. Bump NJ, Hackett M, Hugunin M, Seshagiri S, Brady K, Chen P, Ferenz C, Franklin S, Ghayur T, Li P, Licari P, Mankovich J, Shi L, Greenberg A, Miller LK, Wong WW (1995) Science 269:1885
156. Tewari M, Dixit VM (1995) J Biol Chem 270:3255
157. Kumar S (1997) Cell Death Diff 4:2
158. Chinnaiyan AM, O'Rourke K, Lane BR, Dixit VM (1997) Science 275:1122
159. Wu D, Wallen HD, Nunez G (1997) Science 275:1126
160. Yang J, Liu X, Bhalla K, Kim CN, Ibrado AM, Cai J, Peng TI, Jones DP, Wang X (1997) Science 275:1129
161. Kroemer G, Zamzami N, Susin SA (1997) Immunol Today 18:44
162. Rao L, Debbas M, Sabbatini P, Hockenbery D, Korsmeyer S, White E (1992) Proc Natl Acad Sci USA 89:7742
163. Yew PR, Berk AJ (1992) Nature 357:82
164. Henderson S, Huen D, Rowe C, Dawson G, Johnson G, Rickinson A (1993) Proc Natl Acad Sci USA 90:8479
165. Ray CA, Black RA, Kronheim TA, Greenstreet PR, Sleath GS, Salvesen GS, Pickup DJ (1992) Cell 69:597
166. Clem RJ, Hardwick JM, Miller LK (1996) Cell Death and Diff 3:9
167. Roy N, Mahadevan MS, McLean M, Shutler G, Yaraghi Z, Farahani R, Baird S, Besner-Johnston A, Lefebvre C, Kang X, Salih M, Aubry A, Tamai K, Guan X, Ioannou P, Crawford TO, de Jong PJ, Surh L, Ikeda JE, Korneluk RG, MacKenzie A (1995) Cell 80:167
168. Carson DA, Ribeiro JM (1993) The Lancet 341:1251
169. Gorman AM, McGowan A, O'Neill C, Cotter T (1996) J Neurological Sci 139:45
171. Portera-Cailliau C, Sung C-H, Nathans J, Adler R (1994) Proc Natl Acad Sci USA 91:974
172. Ameisen JC (1995) HIV infection and T-cell death. In Gregory CD (ed) Apoptosis and the immune response. Wiley-Liss, New York, p 115
173. Lacronique V, Mignon A, Fabre M, Violett B, Rouquet N, Molina T, Porteu A, Henrion A, Bouscary D, Varlet P, Joilin V, Kahn A (1996) Nature Med 2:80
174. Nagata S, Suda T (1995) Immunol Today 16:39
175. Rieux-Laucat F, Deist F, Hivroz C, Roberts A, Debatin K, Fischer A, de Villartay J (1995) Science 268:1347
176. Pace GW, Leaf CD (1995) Free Rads Biol Med 19:523
177. Cerrutti PA (1985) Science 227:375
178. Mercille S, Massie B (1994) Biotechnol Bioeng 44:1140
179. Al-Rubeai M, Singh RP, Goldman MH, Emery AN (1995) Biotechnol Bioeng 45:463
180. Simpson NH, Milner AE, Al-Rubeai M (1997) Biotechnol. Bioeng (in press)
181. Xiang J, Chao DT, Korsmeyer SJ (1996) Proc Natl Acad Sci USA 93:14,559

Received October 1997

Measurement of Apoptosis

Zbigniew Darzynkiewicz[1] · Frank Traganos[2]

[1] The Cancer Research Institute, New York Medical College, 100 Grasslands Road, Elmsford, N.Y. 10523, *E-mail: darzynk@nymc.edu*
[2] The Cancer Research Institute, New York Medical College, Valhalla, N.Y. 10595

The cell dying by apoptosis undergoes a sequence of morphological, biochemical, and molecular changes which are characteristic, and often unique, to this mode of cell death. Specific features of apoptotic cells resulting from these changes, which serve as markers used to reveal the apoptotic mode of cell death and to quantify the extent of apoptosis in cultures or in tissue, are reviewed. Analysis of these features by flow or image cytometry is the most commonly used approach to detect, quantify, and study various aspects of apoptosis. Flow or laser scanning cytometry also offer all the advantages of rapid, accurate and multiparametric measurements to investigate the biological processes associated with cell death. Numerous methods have been developed to identify apoptotic and necrotic cells, which are widely used in various disciplines, particularly in oncology and immunology. The methods based on changes in cell morphology, plasma membrane molecular structure and transport function, function of cell organelles, DNA stability to denaturation and endonucleolytic DNA degradation are reviewed and their applicability in the research laboratory and in the clinical setting is discussed. The most common pitfalls and improper use of the methodology in analysis of cell death and in data interpretation are also discussed.

Keywords: Endonuclease, DNA cleavage, Chromatin condensation, Annexin V, Light scatter.

Advances in Biochemical Engineering/
Biotechnology, Vol. 62
Managing Editor: Th. Scheper
© Springer-Verlag Berlin Heidelberg 1998

1
Introduction

1.1
Cell Necrobiology

Mechanisms associated with cell death have become the subject of intense studies during the past decade. The interest in cell death spans a wide range of diverse disciplines, including cell and molecular biology, oncology, immunology, embryology, endocrinology, hematology, neurology, plant biology, and biotechnology. Of particular interest are molecular mechanisms which predispose the cell to respond to an environmental or intrinsic signal by death and/or regulate the initial steps leading to the irreversible commitment to death. The scope of this subject is so wide and involves so many disciplines that it calls for recognition as a separate, new field of biology and medicine. The term "cell necrobiology" has been introduced [1] to define this field. Cell necrobiology, thus, represents a wide area of research which includes different modes of cell death, the biological processes which predispose, precede, and accompany cell death, as well as the consequences and tissue response to cell death. Although the term combines the mutually exclusive elements, *necros* (death) and *bios* (life), it is not contradictory. In preparation for, and during, cell death, a complex cascade of biological processes, typical of cell life, takes place. These processes involve activation of many regulatory pathways, preservation and often modulation of transcriptional and translational activities, alteration in the function of cell organelles, activation of many diverse enzyme systems, modification of the cell plasma membrane structure and function, etc. A term which refers to the biology of cell death, thus, is not a contradiction. On the other hand the term "cell necrology" may be reserved to define specifically the field of studies dealing with the post-mortem changes, occurring in the cell which has passed the point of "no return" in its journey towards death [1].

It is generally accepted that the cell can die by either the mechanism of apoptosis or by necrosis [2]. Apoptosis, frequently defined as "programmed cell death" or "cell suicide", is an active and physiological mode of cell death, in which the cell itself activates and executes the genetic program of its own demise [2–10]. If apoptosis occurs in tissue, the remains of the dead cell are disposed of in such a fashion that the possible adverse consequences for the host such as inflammation or scar formation are minimal. The alternative mode of cell death, necrosis, also defined as "accidental cell death", and with even more literary flavor as "cell murder", is a passive and degenerative process, often induced by massive, suprapharmacological doses of cytotoxic agents. In contrast to apoptosis, necrosis triggers an inflammatory response and often results in tissue scarring. The differences between these modes of cell death, as well as the modes which do not fit the classical picture of either apoptosis or necrosis, are discussed later in this article.

1.2
Multidisciplinary Interest in Cell Necrobiology

A multistep complex mechanism regulates the cell's propensity to respond to various stimuli by apoptosis [11–15]. The regulatory system involves the presence of at least two distinct stages, one controlled by the Bcl-2 family of proteins [13–18], and another by cysteine- [19–22] and also possibly by serine- [23–26] proteases. Among the proteins controlling the first stage, Bcl-2, Mcl-1, and Bcl x_L promote cell survival while Bax, Bcl-x_S, Bak, and Bad enhance the cell's propensity to undergo apoptosis. All members of the Bcl-2 family contain well conserved domains that allow for the formation of homo- and hetero-dimers between members. The cysteine (ICE) proteases, recently renamed caspases [22], consist of at least 11 homologous enzymes with somewhat different substrate specificities.

Caspases interact with the receptors transferring apoptotic signals through the plasma membrane such as Fas or TNF receptor 1 (TNFR1) via adaptor molecules such as FADD/MORT1, through interactions between the well conserved death domains of these proteins [13, 27, 28]. There is a growing body of evidence that the Bcl-2 family of proteins act upstream of caspases [13]. Although not being direct modulators of caspases, the Bcl-2 proteins appear to activate these proteases perhaps by engaging their activator [13, 29].

The wide interest in apoptosis stems from the realization that it is an active and, as mentioned above, highly regulated mode of cell death. In addition to the Bcl-2 family of proteins and caspases, a plethora of other molecules, which either promote or prevent apoptosis, have been discovered to interact with the regulatory machinery of apoptosis. Some are the products of oncogenes or tumor suppressor genes, e.g., such as c-*myc* or wt p53 [30–32]. Others are viral proteins such as BHRF1, a product of the Epstein Barr virus, which is homologous to Bcl-2, or *crmA*, a product of cowpox virus, an inhibitor of caspases [33, 34]. These virally encoded proteins promote their intracellular propagation by postponing apoptosis. Given such a wide range of regulatory steps and molecules, there are many possibilities for interaction with the components involved in the regulation of apoptosis and thereby to modulate the cell's propensity to die. It is not surprising, therefore, that apoptotic mechanisms are the focus of interest of many researchers in the field of oncology. Numerous strategies designed to utilize these mechanisms to modulate the sensitivity of the tumor and/or normal cells to antitumor agents, and, as a result, to increase the efficiency of treatment and/or to lower drug toxicity to the patient, have been considered [35–40].

Another finding promoting interest in apoptosis was the identification of the gene protecting cells from apoptosis (Bcl-2) as an oncogene. This finding revealed that not only a defect in the cell cycle results in uncontrolled cell proliferation but also the loss of the cells' ability to die on schedule may be a cause of cancer. More recently, it became apparent that tumor progression and increased malignancy are also associated with the change in the propensity of tumor cells to undergo spontaneous apoptosis [41–43]. The rate of spontaneous apoptosis in tumors as well as apoptosis induced by chemotherapy, therefore, have become areas of intense interest in oncology. In the latter case, it has recently been demonstrated that the ability of antitumor agents to induce apoptosis can be analyzed in the course of therapy, thereby providing the possibility of rapid assessment of their efficiency [44–47].

Another discipline where wide interest is focused on apoptosis, is immunology. Apoptosis plays a fundamental role in the clonal selection of T cells and is implicated in many other normal and pathological reactions [4, 48]. Of particular interest is the mechanism of cell kill by NK lymphocytes, which is based on the use of the apoptotic effector machinery [49]. Furthermore, since progression of AIDS appears to be correlated with the rate at which T cells die by apoptosis, attempts are being made to monitor apoptosis of these cells and evaluate such measurements as markers of disease progression and prognosis [50, 51].

The essential role of programmed cell death in tissue and organ development was recognized very early by embryologists. Apoptosis plays a role not only in normal tissue and organ modelling during embryogenesis, but is also triggered by some drugs, environmental toxins or other teratogenic agents, where it may be ectopic or unscheduled, leading to congenital malformations.

Male fertility is still another field where apoptosis appears to be of particular interest. It has been observed that DNA in chromatin of abnormal infertile sperm cells, in contrast to normal sperm cells, is excessively sensitive to heat or acid-induced denaturation [52]. The DNA in these cells also has numerous strand breaks [53]. Furthermore, the DNA in situ in sperm is more accessible to actinomycin D [54]. These three features are also typical of DNA in the chromatin of apoptotic somatic cells [55, 56]. Therefore, it was proposed that an apoptosis-like mechanism may be triggered to eliminate cells bearing various types of DNA damage, including mutations, even at late stages of spermatogenesis [43]. However, because such highly differentiated cells may have already lost many effectors of apoptosis, the process may be incomplete, resulting only in activation of an endonuclease which causes extensive DNA degradation, effectively eliminating such cells from the reproductive pool but not leading to their physical disintegration. Such spermatozoa may still have mitochondrial activity, normal motility, and, in some cases, even normal morphology [52]. Analysis of in situ DNA denaturability in sperm cells, assayed by flow cytometry, has become increasingly popular as a marker of infertility and in toxicology studies to assay the genotoxic effects of environmental agents [57, 58].

1.3
Interest in Methodologies of Cell Death Analysis

Because of such widespread interest in necrobiology, numerous methods have been developed to reveal the mode of cell death and to identify and quantify dead cells, in particular cells which died by apoptosis. Initially, apoptotic cells were recognized, based on their characteristic morphology, by light or electron microscopy [2, 3, 7, 9, 10]. Agarose gel electrophoresis of DNA extracted from apoptotic cells, by revealing a characteristic pattern of preferential DNA cleavage at internucleosomal sections ("DNA laddering"), a hallmark of apoptosis [2, 5], has become a common method used to confirm the apoptotic mechanism of cell death.

The flow cytometer has recently become the instrument of choice for analysis of apoptosis in a variety of diverse cell systems (reviews: [1, 59–64]). Cytometry offers the possibility of rapid and very accurate analysis of large populations of individual cells [65, 66]. Selection of cells to analyse is unbiased by the investigator's choice. The most attractive feature of cytometry is the possibility of simultaneous measurement of several parameters on the same cell populations. Such measurements directly reveal correlations between the measured parameters on a cell by cell basis. A variety of cytometric assays have been developed and new assays and modifications of established methods are being introduced at a rapid pace. Several method chapters on the use of cytometry to analyze the modes of cell death have recently been published [61–64]. The present paper reviews the most widely used methodologies in the field of cell necrobiology and updates the earlier reviews on this subject [1, 59, 60].

2
Characteristic Features of Cells Dying by Apoptosis or Necrosis Providing Markers for their Identification

2.1
Changes in Cell Morphology During Apoptosis

Table 1 lists the morphological and molecular changes which occur during apoptosis. As mentioned, many of these changes serve as markers to identify apoptotic cells by microscopy or using flow or laser scanning cytometry. One of the early events of apoptosis is cell dehydration. Loss of intracellular water leads to condensation of the cytoplasm followed by a change in cell shape and size. The cells, originally round, may become elongated or acquire irregular shape and become smaller. Another change is condensation of nuclear chromatin.

Table 1. Changes in cell morphology and molecular events during apoptosis

---

A. Morphological Changes

- Cell shrinkage
- Cell shape change
- Nuclear chromatin condensation
    - Loss of visually recognizable nuclear structure (framework)
    - DNA hyperchromicity
- Dissolution of nuclear envelope
- Nuclear fragmentation
- Condensation of cytoplasm
- Loss of cell surface structures (pseudopodia, microvilli)
- Formation of apoptotic bodies ("budding", "blebbing")
- Detachment of cells in cultures
- Phagocytosis of the apoptotic cell remains

B. Biochemical and Molecular Events of Apoptosis

- Increased ratio of apoptosis promoters vs inhibitors of the Bcl-2 family
- Mitochondrial changes
    - Decrease in the transmembrane potential
    - Leakage of cytochrome C
    - Oxidative stress (formation of ROI)
- Intracellular Ca^{+2} rise
- Cell dehydration
- Loss of asymmetry in plasma membrane phospholipids
- Activation of serine protease(s)
- Cascade-activation of caspases
    - Proteolysis of the "death substrates"
- Degradation of F actin and proteins other than "death substrates"
- Loss of DNA stability to denaturation
- Presence of ss DNA sections
- Endonucleolytic DNA degradation
    - 50–300 kb fragments
    - Cleavage in internucleosomal DNA
- Activation of transglutaminase

Chromatin condensation is one of the most characteristic features of apoptosis. The condensation starts at the periphery of the nucleus with the areas of condensed chromatin often acquiring a concave shape resembling a half-moon or horseshoe. The condensed chromatin has a very uniform, smooth appearance, with no evidence of any texture normally seen in the interphase nucleus. DNA in highly condensed (pycnotic) chromatin exhibits hyperchromasia, staining strongly with fluorescent or light absorbing dyes (Fig. 1).

During further progression of apoptosis the nuclear envelope disintegrates, lamin proteins are proteolytically degraded, and nucleus undergoes fragmentation (karyorrhexis). Nuclear fragments, which stain strongly and uniformly with DNA dyes and resemble heterogenous size DNA droplets, are then scattered throughout the cytoplasm (Fig. 1). Together with constituents of the cytoplasm (including intact organelles), the nuclear fragments are then packaged and enveloped by fragments of the plasma membrane. These structures, called apoptotic bodies subsequently detach from the surface of the dying cell by the process often defined as "blebbing", which rather resembles the mechanism of "budding" in yeast [2]. Loss of cell surface structures such as pseudopodia or microtubules is also a characteristic feature of apoptosis [67]. When the cells grow attached to flasks in tissue culture they detach during apoptosis and float

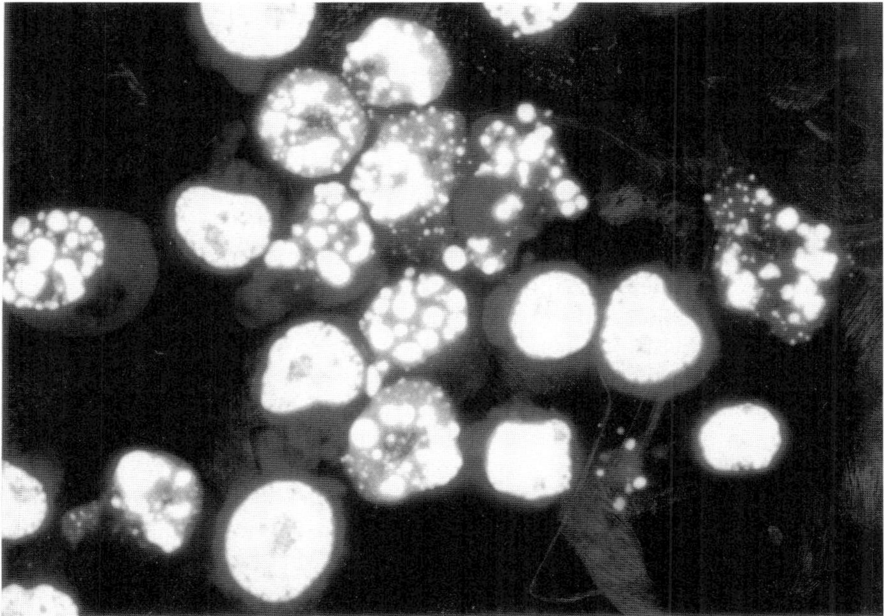

Fig. 1. Detection of apoptotic cells following staining with DAPI and sulforhodamine. HL-60 cells were induced to undergo apoptosis by treatment with 0.15 µmol/l CAM for 4 h (152). The cytospun cells were fixed for 15 min in 1% formaldehyde then in 70% ethanol, stained with 1 µg/ml of DAPI and 20 µg/ml sulforhodamine 101 in PBS and viewed under UV light illumination. Note the typical appearance of apoptotic cells having fragmented nuclei and the presence of individual apoptotic bodies (Adapted with permission from [1])

in the medium. Therefore, discarding media and collecting only the trypsinized cells results in preferential loss of apoptotic cells. When apoptosis occurs, in vivo apoptotic bodies are phagocytized by neighboring cells, including those of epithelial or fibroblast origin (i.e., not necessarily by "professional" macrophages), without triggering an inflammatory reaction or scar formation in the tissue [2, 3, 7, 9, 10].

2.2
Biochemical and Molecular Events of Apoptosis

Changes in the ratio of apoptosis promoters of the Bcl-2 protein family, such as Bax, Bik, Bad, and Bcl-x_S, over inhibitors of apoptosis such as Bcl-2, Bcl-x_L, A-1, Bcl-w, and Mcl-1, appears to be among the earliest molecular events of apoptosis [13–15, 68]. These proteins dimerize with one another and their propensity for dimerization is modulated by their phosphorylation [69]. Little is known about the molecular mechanism by which the Bcl-2 proteins execute their function in promoting or inhibiting apoptosis. Some are localized on the outer mitochondrial membrane, nuclear envelope, and endoplasmic reticulum. The structure of Bcl-x_L [70] suggests that this protein may form pores in biological membranes and that interactions between the antagonistic members of the Bcl-2 family may regulate the pore's permeability. The loss of mitochondrial transmembrane potential observed early during apoptosis [71–75], appearance of reactive oxidative intermediates (ROIs) within the cell [76], the release of cytochrome C [77, 78] as well as Ca^{2+} from mitochondria [79], all are presumed to be a consequence of the opening of pores in the mitochondrial membrane [80]. Release of Ca^{2+} causes a rise in intracellular concentration of this cation [79]. These events are upstream of activation of caspases [29, 81, 82], and appear to be the earliest events, perhaps the ones actually triggering the cascade of apoptotic processes, ultimately leading to cell disintegration.

The structural integrity and most of the plasma membrane transport function is preserved at least during the initial phase of apoptosis. However, plasma membrane permability to certain fluorochromes, such as 7-AAD, Hoechst 33342, or Hoechst 33258, is increased [83–85]. The most characteristic change in the plasma membrane, however, is the loss of asymmetry of the phospholipids on the plasma membrane leading to exposure of phosphatidylserine on the outer surface [86–89]. This change occurs early during apoptosis regardless of whether apoptosis is induced by agents directly interacting with receptors on the plasma membrane such as the Fas receptor, or by genotoxic agents, triggering cell death via DNA damage. Exposure of phosphatidylserine on the outer leaflet of the plasma membrane preconditions remnants of the apoptotic cell (apoptotic bodies) to become a target for phagocyting cells. Loss of pseudopodia or microvilli is paralleled by degradation of F actin [67]. Still another event of apoptosis is activation of transglutaminase which crosslinks cytoplasmic proteins [89] resulting in a change in the physical properties of the cell, making them stiffer and resistant to deformation under pressure.

The cascade-like activation of serine proteases and caspases, mentioned in the Introduction, leads to proteolytic degradation of selected proteins, defined as "death substrates". Poly(ADP)ribose polymerase (PARP), lamin, actin, U1 small nuclear ribonucleoprotein (U1 snrp), rho-DGI, SREBP, and DNA-dependent protein kinase have been recognized among these substrates [19–21, 81, 90]. Involvement of serine proteases was observed in the specific degradation of two nuclear proteins, lamin B and NuMA [26]. Some specificity in degradation of the proliferation-associated nuclear proteins Ki-67, p120, and proliferating cell nuclear antigen (PCNA) has also been observed [91].

The activation of an endonuclease(s) is another very characteristic event of apoptosis [2, 3, 5, 10]. Initially, DNA is cleaved at the sites of attachment of chromatin loops to the nuclear matrix, which results in the appearance of discrete 300–50 kb size fragments [92]. Subsequently, DNA is preferentially cleaved between nucleosomes. The products are discontinuous DNA fragments representing nucleosomal- and oligonucleosomal-sized DNA sections. They generate a characteristic "ladder" pattern during agarose gel electrophoresis. Because DNA in apoptotic cells is partially degraded, the fraction of low molecular weight DNA can be extracted from these cells, following either permeabilization by detergents, or prefixation in precipitating fixatives such as alcohols or acetone [3,5,93]. It should be pointed out, however, that in many cell types, DNA cleavage during apoptosis does not proceed to internucleosomal-sized sections but rather proceeds only to 300–50 kb size DNA fragments [92, 94–98].

2.3
Duration of Apoptosis. Method-Dependent Differences in the Time Windows for Detection of Apoptosis

The duration of apoptosis, from the onset of the event to total disintegration and disappearance of the cell, may vary. Generally, within live tissue, apoptosis is of short duration, frequently being shorter in length than mitosis [2, 9, 10]. Rapid disappearance of apoptotic cells is apparently facilitated by their phagocytosis by immediately adjacent cells. Thus, under conditions of cell homeostasis in normal tissue, when the rate of cell death is balanced by the rate of cell proliferation, the mitotic index may exceed the index of apoptosis. Because of their low frequency, it is difficult to find apoptotic cells under these conditions. In tissue culture, however, apoptotic cells may remain for extended periods of time, enduring for many hours before their disintegration.

Identification of apoptotic cells generally relies on a particular characteristic feature of apoptosis discussed above. The time period during which the cells undergoing apoptosis present such a feature varies, depending on the feature. Some events are very transient, while others last longer. Altered permeability of the plasma membrane to 7-AAD, for example, is a very early event of apoptosis and it occurs before the appearance of DNA strand breaks. Later in the apoptotic process, the cells can be recognized based on the presence of DNA strand breaks. At this point, however, uptake of 7-AAD is useless for the identification of apoptotic cells because the fluorochrome enters as rapidly into necrotic cells. Thus, because the time window for identification of apoptotic cells varies depending on the method, the frequency of apoptotic cells (apoptotic index) may be different, depending on the assay which is used.

2.4
Accidental Cell Death (Necrosis) vs Typical and Atypical Apoptosis

While apoptosis is characterized by the active participation of the affected cell in its own destruction, even to the point of triggering (in some cell systems), the de novo synthesis of cell death effector molecules, necrosis, generally results from the gross injury to the cell induced by an overdose of cytotoxic agents and is a passive, catabolic, and degenerative process. An early event of necrosis is mitochondrial swelling, which is followed by rupture of the plasma membrane and release of cytoplasmic constituents, including proteolytic enzymes, outside of the cell [2, 9]. Nuclear chromatin shows patchy areas of condensation and the nucleus undergoes slow dissolution (karyolysis). Necrosis triggers an inflammatory reaction in the tissue and often results in scar formation. DNA degradation is much less extensive during necrosis compared to apoptosis [24], and the products of degradation are heterogenous in size, forming a smear rather than discrete bands on electrophoretic gels.

Cell death does not always manifest classical features of either apoptosis or necrosis. Many examples of cell death have been described in which the pattern of morphological and/or biochemical changes resemble neither typical apoptosis nor necrosis but often contain features of both [94–101]. In some cases, the integrity of the plasma membrane was preserved but DNA degradation was random, without evidence of internucleosomal cleavage. In other situations, DNA degradation was typical of apoptosis but nuclear fragmentation and other features of apoptosis were not apparent. Generally, while most hematopoietic lineage cell types are "primed" to undergo apoptosis and their death has typical features of apoptosis, the death of epithelial type cells or fibroblasts provide a more complex pattern that is sometimes difficult to classify. Furthermore, some drugs which cause apoptosis may additionally confuse the pattern of cell death due to drug-induced secondary effects on the cell. When apoptosis is triggered by drugs affecting cell structure and function, or by drugs affecting one or more pathways of the apoptotic cascade, particular features of apoptosis may not be apparent. Induction of apoptosis in the presence of inhibitors of proteases, for example, can lead to cell death without evidence of nuclear fragmentation or a decrease in the stability of DNA to denaturation [102]. Likewise, prolonged cell arrest in the cell cycle induced by some drugs can result in growth imbalance which may dramatically alter cell biochemistry and morphology [103]. The apoptotic features of such cells are much different compared to untreated cells.

The complexity in defining the mode of cell death and in classification of different death mechanisms was recently reviewed by Majno and Joris [2]. These authors convincingly argue that cell death and necrosis are two distinct events. Cell death, according to these authors, is a process that leads to the point of no return; subsequent events are the post-mortem changes, which these authors define as necrotic changes. In the case of induction of liver cell death by ischemia, for example, the irreversible point is at approximately 150 min after onset of oxygen deprivation [2]. At that time, however, no significant morphological changes can be seen, while the necrotic changes become visible only after 12 h.

The cells die, thus, long before any morphological changes typical of necrosis can be detected. The term cell death "by necrosis", which implies that necrotic changes accompany cell death, appears to be contradictory and confusing. Additionally confusing is the traditional use of the term "necrosis" in pathology to describe gross tissue changes, visible by eye and occurring after cell death, i.e., attributed to postmortem cell changes.

The term "accidental cell death" to define the mode of cell death which is now generally denoted as cell necrosis, and "oncosis" to portray the early stages of accidental cell death was proposed by Majno and Joris [2]. The term oncosis has been used by pathologists to describe cell death associated with cell swelling (*óncos* = swelling) which occurs during slow ischemia, e.g., as in the case of death of osteocytes entombed in bone tissue during osteogenesis. The features of oncosis are identical to those seen in the early phase of accidental cell death in a variety of cell systems. Within the framework of classification of cell death proposed by Majno and Joris [2], the necrotic step follows either oncosis or apoptosis and thus can be denoted either as "oncotic necrosis" or "apoptotic necrosis", respectively. The common features of such late necrotic cells, regardless of whether they were dying by apoptosis or oncosis, are loss of plasma membrane transport (ability to exclude charged dyes such as PI or trypan blue), autolytic processes, dissolution of the remnants of chromatin (karyolysis), etc. According to the terminology which we recently proposed [1] the term "cell necrobiology" may refer to the studies of the events which occur during both apoptosis and oncosis, while "cell necrology" refers to the post mortem events, defined by Majno and Joris as "necrosis". The taxonomy of cell death proposed by these authors [2] is rational and has attractive elements which may clarify some inconsistencies in the current classification as discussed above.

2.5
Early and Delayed Apoptosis. Relationship to the Cell Cycle

It has been observed in numerous studies that exposure of lymphocytes or hematological tumor cell lines to pharmacological concentrations of antitumor drugs or radiation induces their rapid apoptosis, which can be seen as early as 2–4 h after the onset of treatment [104, 105]. Likewise, apoptosis of lymphocytes induced by their activation, corticosteroids, or via engagement of the Fas receptor, by Fas ligand, or by TNF, also is rapid. In all these cases, the pattern of apoptosis is quite typical and the cells exhibit nearly all the characteristic apoptotic features, as listed in Table 1. During this early apoptosis the cells generally die in the same cell cycle, prior to mitosis. A term "homo-cycle" apoptosis has been proposed to define apoptosis which occurs in the same cycle in which the cells were initially exposed to the apoptosis-triggering agent, prior to, or during mitosis. Often the cells die in the same phase of the cycle, without progressing to the next one. To define such death, the term "homo-phase apoptosis" applies [47]. An example of such apoptosis is rapid death of S phase cells during treatment of exponentially growing cell populations with DNA topoisomerase inhibitors [105, 106]. In contrast to "homo-cycle apoptosis", the term "postmitotic apoptosis" has been proposed [47] to describe apoptosis which occurs in the cell cycle(s) subsequent to the one in which the initial damage to the cell was induced.

This subdivision is pertinent in relation to the possible differences in mechanisms triggering apoptosis. During homo-phase apoptosis, which, for example, occurs in S phase, the cells do not traverse G_1 or G_2 checkpoints. Therefore the product of the tumor suppressor gene p53, which controls these checkpoints, may not play a significant sensitizing role, as it otherwise does in cells that traverse these checkpoints. Antitumor strategies involving p53 and the checkpoints, therefore, may be different for drugs inducing homo-phase vs postmitotic apoptosis.

While early apoptosis occurs generally between 3 and 8 h after induction of damage in the cell, apoptosis may also occur after a considerable delay. For example, exposure of cells to antitumor drugs at low concentrations or when transient (pulse) treatment with drugs or radiation is followed by growth in drug-free media, the initial results may be perturbation of cell cycle progression (cytostatic effect). Subsequently, such cells may either resume cell cycle progression and even be capable of reproduction after removal of the drug from the medium, or they may die with a delay. Cell death in such a situation is often due not only to the primary damage but may also result from the accumulation of secondary changes which occurred during the cytostatic phase. One such change is growth imbalance. During inhibition of DNA replication, RNA and protein synthesis continue, but many genes which were initially damaged by the drug may become transcriptionally inactive. The cells die if products of these genes are essential for their survival. Less is known about whether altered ratios of overall RNA/DNA or protein/DNA, gross features characteristic of cell growth imbalance may themselves contribute to cell death.

Delayed apoptosis is also observed in the case of certain cell types, in particular cells of epithelial origin even following exposure to cytotoxic drug concentrations or radiation [e.g., 107, 108]. It is possible that the cells which do not undergo early apoptosis lack the required effectors of apoptosis. The apoptosis triggering event, in these cells, induces transcription of genes encoding the effector(s), which explains a delay in the execution of apoptosis. Inhibition of protein synthesis by cycloheximide, in such cases, generally protects the cells from apoptosis. For example, human T cell leukemic MOLT-4 cells do not undergo early apoptosis [109], most likely because they lack the action of the apoptosis-associated endonuclease, which is induced with a delay [110].

Interestingly, the phase of the cell cycle in which the cells undergo apoptosis in response to a particular drug is often different in the case of early vs delayed apoptosis. Transient exposure of cells to a drug which, for example, induces early apoptosis in S (e.g., CPT or teniposide) or in G_2 M phase (taxol), followed by cell growth in drug-free medium, leads to cell death not only in S or G_2 M but often in G_1 phase [47].

3
Assays of Apoptosis not Utilizing Flow Cytometry

3.1
Analysis of Cell Morphology

Apoptosis was originally recognized based on typical changes in cell morphology as revealed by light or electron microscopy [2, 7, 9, 10], and reviewed earlier in this article. In most cell types the morphological features of cells dying by apoptosis are very characteristic and thereby their identification based on morphological criteria is simple. For light microscopy, approximately $1-2 \times 10^5$ cells suspended in 100–200 µl of a solution of PBS containing 1 % (w/v) bovine serum albumin (BSA) should be cytocentrifuged on microscopic slides using a rather low speed of the cytospin centrifuge and short centrifugation time (e.g. 600–1000 rpm for 2–4 min). Because apoptotic cells are fragile, it is advisable to precoat the microscope slides with BSA prior to using it for depositing cells, by immersing the slides in a 1 % (w/v) solution of BSA in water and allowing the slide to dry. After cytocentrifugation, the slides should be air-dried, then fixed for 15 min in a 1 % formaldehyde solution on ice, and transferred and stored in 80 % ethanol. A variety of stains can be used to visualize the cells, including the routine hematoxylin and eosin combination or 0.1 % Meyer's hematoxylin alone for transmission light microscopy, or DNA nucleic acid reactive fluorochromes such as AO, Hoechst 33342 or 33258 dyes, DAPI, PI or 7-AAD for UV light fluorescence microscopy. Details of the methods applied to cytospun cells or tissue sections are presented elsewhere [111, 112].

The most commonly used morphological criteria for identification of apoptotic or necrotic cells are listed in Table 2. These changes were discussed earlier.

Table 2. Morphological criteria for identification of apoptosis or necrosis

Apoptosis

- Reduced cell size, convoluted cell shape
- Plasma membrane undulations ("blebbing/budding")
- Chromatin condensation (DNA hyperchromicity)
- Loss of the structural features of the nucleus (smooth appearance)
- Nuclear fragmentation (karyorrhexis)
- Presence of apoptotic bodies
- Dilation of the endoplasmic reticulum
- Relatively unchanged cell organelles
- Phagocytosis of cell remnants
- Cell detachment from tissue culture flasks

Necrosis

- Cell and nuclear swelling
- Patchy chromatin condensation
- Swelling of mitochondria
- Vacuolization in cytoplasm
- Plasma membrane rupture ("ghost-like" cells)
- Dissolution of DNA (karyolysis)
- Attraction of inflammatory cells

Electron microscopy is used to reveal ultrastructural changes during apoptosis. The cell preparation and methodology (cell or tissue fixation, embedding, sectioning, etc.) for identification of apoptotic cells is routine and does not require unusual procedures. Transmission electron microscopy reveals ultrastructural changes such as chromatin condensation, loss of the nuclear envelope, dilatation of the endoplasmic reticulum, relatively minor alterations in cytoplasmic and cell organelle structure, all of which are considered markers of apoptosis. Scanning electron microscopy reveals plasma membrane changes such as loss of microvilli and pseudopodia and the presence of cell surface undulations ("blebbing – budding") typical of apoptosis. Because electron microscopy is very tedious, it is rarely used, except to evaluate the mode of cell death in atypical cases or to reveal unusual structural changes during apoptosis or necrosis.

Time lapse microcinematography is another methodology based on analysis of changes in cell morphology, which is occasionally used to detect and even quantify apoptosis. It is applicable only to cells growing attached to flasks. Their detachment, correlated with the observation of undulations to the plasma membrane, recorded by microcinematography, provide sufficient evidence of the apoptotic mode of cell death.

The major advantage of morphological assays is that identification of apoptotic cells is based on the classical criteria of apoptosis [2, 7, 9, 10]. In doubtful cases, therefore, identification of apoptosis by morphological criteria should be used, overriding identification by other methods. The disadvantage of these assays is that the methodology is cumbersome and selection of the cells to be analyzed is not random but subjective and possibly biased.

3.2
Analysis of DNA Fragmentation

As mentioned, activation of endonuclease(s) which cleaves DNA into discrete fragments, initially into 300–50 kb- and subsequently into 180 bp-sections [92] is considered to be a hallmark of apoptosis. This event, therefore, provides a marker for identification of apoptotic cells and measurement of the extent of apoptosis. There are several different approaches to detect DNA degradation, as listed in Table 3. Some of these methods are associated with flow cytometric cell analysis and will be described later.

Table 3. Analysis of DNA fragmentation during apoptosis

- Agarose gel electrophoresis
  - Detection of 180 bp DNA fragments
  - Detection of 300–50 kb DNA fragments (FIGE)
- Individual cell electrophoresis ("comets" assay)
- Detection of low MW (extractable) DNA
- Immunochemical detection of nucleosomes
- Detection of cells with fractional DNA content ("sub-G_1" cells)
- In situ DNA strand break labeling
  - DNA polymerase assay
  - TdT assay

Detection of DNA fragments by gel electrophoresis is a widely used methodology to determine the mode of cell death. This analysis, although in essence qualitative, is actually the most common biochemical assay for detection of apoptosis.

DNA is extracted by standard procedures and assayed on 1.5% agarose gels. Detection of 180 bp DNA sections and its multiples, representing DNA of the size of individual nucleosomes and oligonucleosomes is rather straightforward. These products form a characteristic "ladder" pattern after DNA is counterstained with ethidium bromide (Fig. 2).

A selective method of isolation of only low molecular DNA from apoptotic cells, which is very simple and highly sensitive in detecting apoptosis, was recently described [93]. In this method the cells are initially fixed with ethanol, subsequently centrifuged, and the cell pellet resuspended in a small volume of a buffer of high ionic strength. Under these conditions the degraded, low MW DNA is extracted from the pre-fixed cells into the buffer solution while high MW DNA still remains within the cells. The extract is then directly processed by brief pretreatment with RNase and proteinase K, and without additional steps loaded onto the agarose gel for electrophoresis. The very same cells from which DNA has been extracted can be analyzed by flow cytometry and the cells with fractional DNA content are then identified as apoptotic cells [93; see later]. The major advantage of this method is its simplicity and a possibility of analysis of the same cell populations by two independent methods, gel electrophoresis and flow cytometry (Fig. 2).

As mentioned, while DNA cleavage resulting in large fragments of the nucleic acid, between 300 and 50 kb, appears to be ubiquitous during apoptosis, the degradation does not always progress to the internucleosomal DNA sections. The large DNA fragments can be detected by pulsed-field gel electrophoresis techniques such as field-inversion gel electrophoresis (FIGE) [93]. Relatively well-defined bands are then detected at that range of MW, which allows one to distinguish apoptosis from necrosis, during which no discrete DNA fragments are apparent. In the absence of "DNA laddering", therefore, it is advisable to subject DNA to pulsed field gel electrophoresis in order to confirm or exclude the apoptotic mode of cell death.

Individual cells can also be subjected to gel electrophoresis. Toward this end the cells are embedded in thin layer of agarose gel (e.g., placed on the microscope slide) and subjected to electrophoresis. The degraded DNA pulled out by electric field from apoptotic cells is then counterstained with the DNA fluorochromes such as DAPI or PI. The apoptotic cells appear as "comets" having characteristic "heads", which represent the cells' remains containing high MW DNA and "tails", which represent a fraction of the degraded DNA (Fig. 3).

DNA degradation can also be measured immunochemically by using antibodies against DNA and histones. Such antibodies react with DNA-histone complexes, detecting mono- and oligonucleosomes. Several types of cell viability kits based on immunochemical detection of nucleosomes are provided by some reagent companies (e.g., Boehringer Mannheim).

DNA cleavage during apoptosis can also be detected by in situ labeling of DNA strand breaks in fixed and permeabilized cells with either enzyme-(phos-

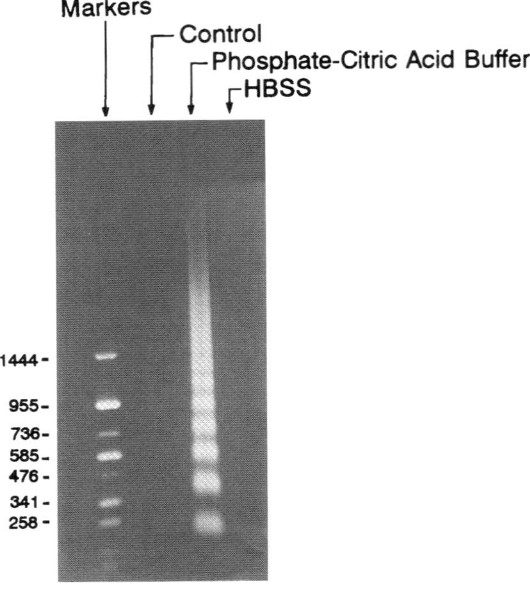

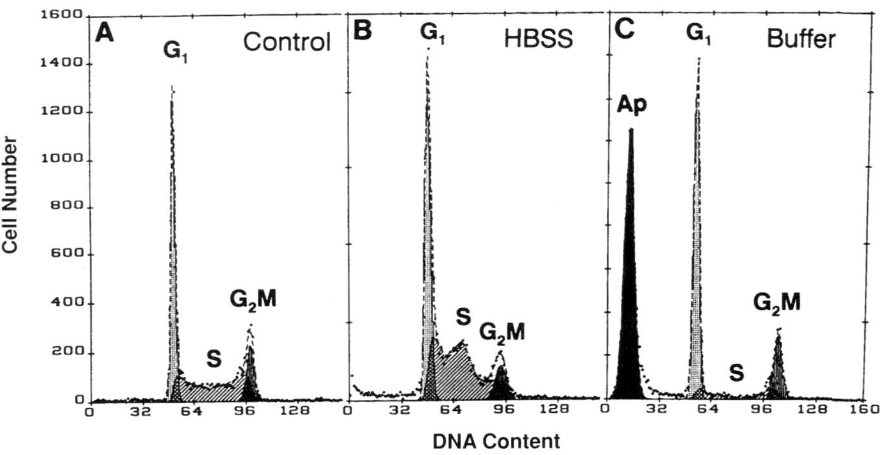

Fig. 2. Combined analysis of apoptosis by DNA gel electrophoresis and flow cytometry of the same cells. DNA gel electrophoresis and DNA content frequency histograms of corresponding HL-60 cells untreated (Control, A) and treated for 4 h with 0.15 µmol/l DNA topoisomerase I inhibitor camptothecin which selectively induces apoptosis of S phase cells (B, C). The cells represented by histogram B were extracted with Hanks' balanced salt solution (HBSS) while these represented by histogram C were extracted with 0.2 mol/l phosphate-citrate buffer at pH 7.8. DNA extracted with HBSS or with the phosphate-citrate buffer was analyzed by gel electrophoresis, as shown in respective lanes, while the same cells after the extraction were stained with DAPI and subjected to flow cytometry. Note the presence of degraded DNA extracted from the camptothecin treated cells by the buffer and its absence in the HBSS extract and in the extract from untreated (control) cells. The DNA marker size is expressed in base pairs (Reproduced with permission from [93])

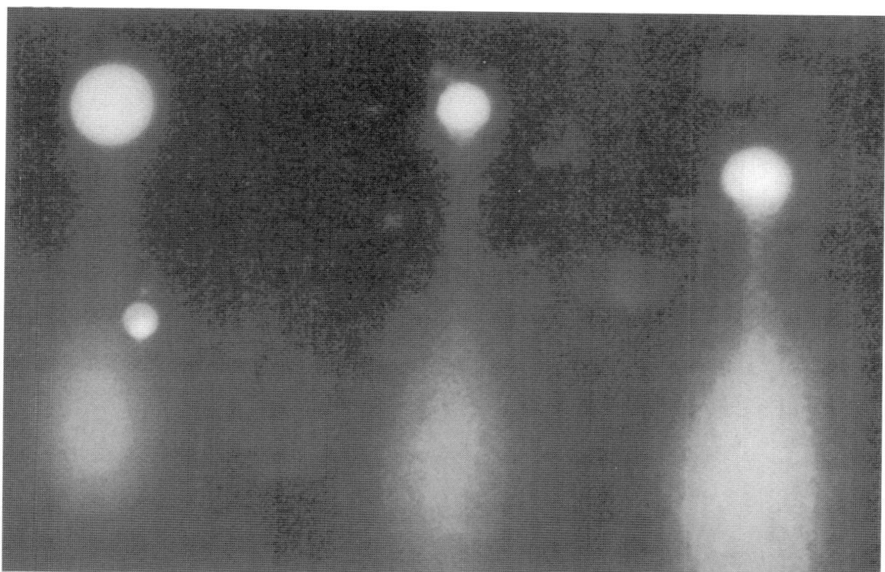

Fig. 3. The "comet" assay of apoptosis. To induce apoptosis exponentially growing HL-60 cells were incubated with 0.15 μmol/l DNA topoisomerase I inhibitor camptothecin for 4 h, fixed in 70 % ethanol, then treated with RNase A and suspended in 1 % agarose film on the microscope slide and subjected to 5 min electrophoresis under conditions as described in [1]. The gels were counterstained with 1 μg/ml of PI and viewed under the fluorescence microscope (green light excitation). Note the typical "comet" appearance of DNA from single cells, with the migrating low MW DNA fraction forming the comet's "tail". The integrated values of PI fluorescence in the "tail" regions represent a fraction of the degraded DNA while the total PI fluorescence ("tail" plus "head") represent total DNA content, correlating with the cell cycle position. (Adapted with permission from [1])

phatase or peroxidase) or fluorochrome-tagged deoxynucleotides using exogenous DNA polymerase or terminal deoxynucleotidyl transferase (TdT) [104, 105, 113 – 115]. Apoptotic cells are then detected by light absorption microscopy, after developing the enzymatic reaction with a colorogenic phosphatase or peroxidase substrate [115, 116], or by UV light microscopy/flow cytometry, when using a fluorochrome-labeled deoxynucleotide [104, 105]. The DNA strand break labeling assays are described further later, under flow and laser scanning cytometry methods.

4
Methods Relying on Flow or Laser Scanning Cytometry

4.1
Analysis of the Laser Light Scatter

Light is scattered when the cell moves through the focus of a laser beam in the flow cytometer. Analysis of the scattered light at different angles provides information about the cell's size, shape, and structure [85]. The intensity of light scattered in the forward direction correlates with cell size. The larger the cell, the more light is scattered. The intensity of scattered light measured at right angles to the laser beam (side scatter), on the other hand, correlates with granularity, refractiveness, and the presence of intracellular structures that can reflect the light [117]. The cell's ability to scatter light is expected to be altered during cell death, reflecting the morphological changes such as cell swelling or shrinkage, breakage of the plasma membrane, and, in the case of apoptosis, cytoplasmic, and chromatin condensation, nuclear fragmentation and shedding of apoptotic bodies.

A decrease in forward light scatter, reflecting cell shrinkage, is observed early during apoptosis [118]. Initially, this decrease is not paralleled by a decrease in side scatter (Fig. 4). The intensity of right angle light scatter initially is either unchanged or, in some cell types, increased [94]. The increase may reflect an increased light reflection and refraction by areas of condensed chromatin and fragmented nuclei. In later stages of apoptosis, however, the intensity of light scattered at both forward and right angle directions is markedly decreased (Fig. 4) [118, 119].

The assay of cell viability by light scatter measurement is simple and can be combined with the analysis of surface immunofluorescence, e. g., to identify the phenotype of the dying cell or detect some other cell marker. It can also be combined with functional assays such as mitochondrial potential, exclusion of PI or plasma membrane permeability to such dyes as Hoechst 33342 or 7-AAD, as will be described further later.

It should be stressed, however, that the light scatter changes are not specific to apoptosis. Mechanically broken cells, isolated cell nuclei, and necrotic cells also have diminished ability to scatter light. Identification of apoptosis by light scatter, therefore, requires additional controls, and should be accompanied by another, more specific assay.

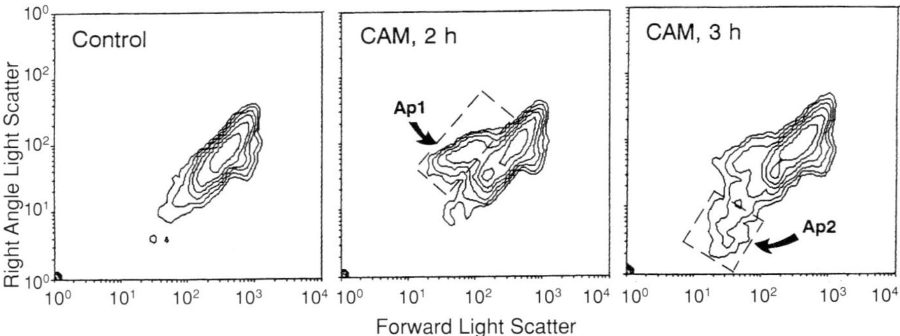

Fig. 4. Changes in the light scatter properties of HL-60 cells treated with 0.15 µmol/l camptothecin for 0, 2 and 3 h. Treatment with camptothecin induced apoptosis in approximately 35 – 40 % of the cells, specifically cells in S phase. The initial decrease in forward light scatter of cells which undergo apoptosis after 2 h of treatment (Ap1) is followed by an even more dramatic change in their ability to scatter light at right angles (Ap2) after 3 h (Reproduced with permission from [59])

In contrast to apoptosis, cell necrosis is initially associated with a transient increase in intensity of light scattered in the forward direction, most likely reflecting cell swelling [2, 9, 10]. A marked decrease in the cell's propensity to scatter light in both the forward and right angle direction follows the initial increase, which in turn is a reflection of the rupture of the plasma membrane and leakage of the cell's constituents.

4.2
Assays Detecting Changes in the Plasma Membrane

4.2.1
Altered Membrane Permeability and Transport Function

The key feature distinguishing dead from live cells is the loss of transport function and often even the loss of structural integrity of the plasma membrane. A variety of assays of cell viability have been developed based on changes in the permeability and transport properties of the plasma membrane. Charged cationic dyes such as trypan blue, propidium, or ethidium, and 7 amino-actinomycin-D (7-AMD) are excluded from live cells with intact plasma membranes. Thus, short incubation with these dyes results in selective labeling of dead cells, while live cells show minimal dye uptake [84, 98, 120–124]. Assays based on exclusion of these fluorochromes are commonly used to probe cell viability. Generally, incubation of cells in the presence of these fluorochromes for 5–10 min labels dead cells, i.e., cells that cannot exclude the dye such as necrotic and late apoptotic cells. The PI exclusion test is frequently used as the flow cytometric equivalent of the trypan blue exclusion assay. These assays can be combined with analysis of cell surface immunophenotype [e.g., 84, 124].

Membrane integrity can also be probed using the nonfluorescent esterase substrate, fluorescein diacetate (FDA). This substrate, after being taken up by live cells, is hydrolyzed by ubiquitous intracellular esterases found in all types of cells [125]. The product of the hydrolysis, fluorescein, is a highly fluorescent, charged molecule which, because of the charge, becomes trapped in intact cells. Incubation of cells in the presence of both PI and FDA thus differentially labels live cells green (fluorescein) and dead cells red (PI). This is a convenient assay, widely used in flow cytometry [61].

The fluorochrome ethidium monoazide (EMA) also is a positively charged molecule similar to EB or PI and is excluded from live and early apoptotic cells. The dye therefore stains only cells which have lost the integrity of their plasma membrane, i.e., necrotic and late apoptotic, as well as mechanically damaged cells [126]. EMA can be photochemically crosslinked to nucleic acids by exposure to visible light. Cell incubation with EMA, followed by their illumination, irreversibly labels the cells which were unable to exclude the dye during incubation. The photolabeling of EMA can be conveniently combined with membrane immunophenotyping [127].

The UV light excited DNA specific fluorochrome Hoechst 33342, unlike PI, is not excluded by live or apoptotic cells. A short exposure of cells to low concentrations of Hoechst 33342 results in strong labeling of apoptotic cells [83, 120, 128–130]. Live cells, on the other hand, require significantly longer incubation with this dye to obtain a comparable intensity of fluorescence. Uptake of Hoechst 33342 combined with exclusion of PI (to identify necrotic and late apoptotic cells) and with analysis of the cell's light scatter properties has been proposed as an assay of apoptosis [83, 120, 130]. Another Hoechst fluorochrome, Hoechst 33258, appears to offer the advantage of increased fluorescence stability compared to Hoechst 33342 [131, 132]. A combination of cell labeling with Hoechst 33342 and 7-AMD was shown to discriminate between apoptotic and necrotic cells and to allow one to reveal the surface immunophenotype [84, 124]. Interestingly, while PI is excluded from both live and apoptotic cells to a similar extent, its rather close analog EB appears to penetrate more efficiently through the plasma membrane of apoptotic cells [133]. The increased stainability of apoptotic cells with EB, however, can also be due to transient chromatin changes occurring early in the apoptotic process [133].

The degree of change in permeability of the plasma membrane to either charged or uncharged fluorochromes varies with the stage (advancement) of apoptosis, cell type, and mode of induction of apoptosis (e.g., DNA damage vs engagement of the Fas receptor). Therefore, the optimal conditions for discrimination of apoptotic cells (fluorochrome concentration, time and temperature of incubation, ionic composition of the incubation medium) may vary significantly between different cell systems. Pilot experiments, therefore, are always necessary to customize the conditions for different cell systems for maximal discrimination of apoptotic from live, and/or from necrotic cells.

4.2.2
Loss of Asymmetry of Plasma Membrane Phospholipids

The plasma membrane phospholipids are asymmetrically distributed between inner and outer leaflets of the plasma membrane of live cells. Specifically, phosphatidylcholine and sphingomyelin are exposed on the external surface of the lipid bilayer while phosphatidylserine is located on the inner surface [86]. It has been shown recently that loss of phospholipid asymmetry, leading to exposure of phosphatidylserine on the outside of the plasma membrane, is an early event of apoptosis [86, 87]. The anticoagulant annexin V preferentially binds to negatively charged phospholipids such as phosphatidylserine. By conjugating fluorescein to annexin V it has been possible to use such a marker to identify apoptotic cells by flow cytometry [87]. During apoptosis the cells become reactive with annexin V after the onset of chromatin condensation but prior to the loss of the plasma membrane's ability to exclude PI. Therefore, by staining cells with a combination of fluoresceinated annexin V and PI it is possible to detect nonapoptotic live cells (annexin V negative/PI negative), early apoptotic cells (annexin V positive, PI negative) and late apoptotic or necrotic cells (PI positive) by flow cytometry (Fig. 5) [87, 134].

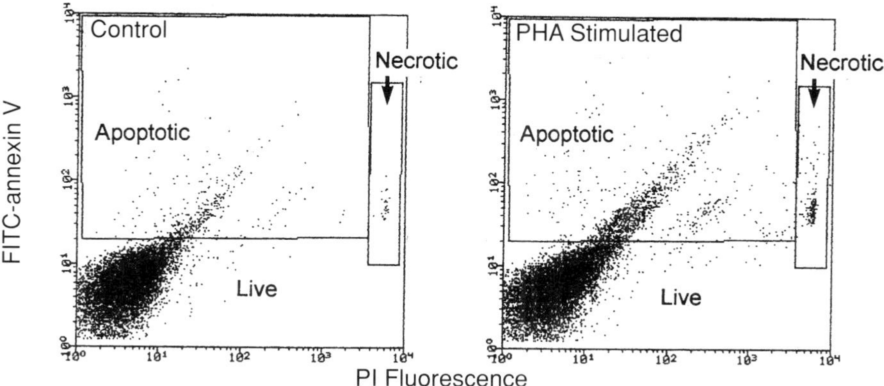

Fig. 5. Supravital staining of apoptotic cells with an FITC-annexin V conjugate. Human peripheral blood mononuclear cells, separated by density gradient centrifugation, were cultured in the absence (*left*) and presence (*right*) of the mitogen phytohemagglutinin (PHA) for 6 h. The majority of cells in both cultures remain viable (live) and thus have only background levels of annexin V-FITC and PI stainability. Cells which have become damaged either as a result of the treatment or during processing as well as late apoptotic cells have dramatically increased PI staining (necrotic). Cells which respond by undergoing "activation-induced" apoptosis show an increase in annexin V-FITC staining and an intermediate PI stainability (apoptotic); this population is clearly increased in the PHA-treated culture (Reproduced with permission from [64])

4.2.3
Other Changes in the Plasma Membrane

A rapid loss of plasma membrane structures such as pseudopodia and microvilli, resulting in the smooth appearance of the cell surface under the electron or phase contrast microscope, characterizes changes that occur relatively early during apoptosis [2, 7, 9, 10]. F-actin is a major constituent of pseudopodia. The phallotoxins are toxic cyclic peptides which bind to F-actin and prevent its depolymerization. Fluoresceinated phallotoxins are used as a probe of F-actin. It has recently been reported that the ability of cells to bind fluoresceinated phalloidin is lost during apoptosis [67]. It was proposed, therefore, to combine cell staining using fluoresceinated phalloidin with DNA content analysis to identify apoptotic cells and to reveal the cell cycle position of both the apoptotic and nonapoptotic cell population [67]. This approach was tested on HL-60 cells induced to undergo apoptosis by etoposide. This is a relatively simple and inexpensive method which, if confirmed on other cell systems, may find many applications.

4.3
Assays of Cell Organelles

4.3.1
Mitochondria

Functional assays of cell organelles provide information about the mechanism of cell death. An early apoptotic event involves a decrease in mitochondrial transmembrane potential [13 – 15, 68]. There is a growing body of evidence that the anti-apoptotic and pro-apoptotic members of the Bcl-2 family of proteins form complexes within the outer mitochondrial membrane and that the initiation of apoptosis is triggered by the imbalance between these pro- vs anti-apoptotic molecules [70 – 73]. The imbalance leads to an opening of the pores across the mitochondrial membrane (presumed to be formed by the pro-apoptotic molecules of the Bcl-2 family), release of cytochrome c from mitochondria, and generation of reactive oxygen intermediates (ROIs). This, in turn, triggers the cascade-like activation of caspases [77, 78]. The loss of the mitochondrial transmembrane potential is apparently a reflection of the opening of the transmembrane pores, which otherwise are protected by the anti-apoptotic molecules of the Bcl-2 family of proteins.

Several different fluorochromes are used as probes of mitochondrial transmembrane potential. The charged cationic green fluorochrome Rh 123 has been frequently used during the past decade to assay the functional state of mitochondria [135, 136]. Cell incubation with Rh 123 results in labeling of live cells while dead cells, having uncharged mitochondria, show minimal Rh 123 retention. Cell incubation with both Rh 123 and PI labels live cells green (Rh 123) and dead cells red [136]. A transient phase of cell death, most likely by necrosis, however, can be detected, when the cells partially lose the ability to exclude PI and yet stain even more intensively than intact cells with Rh 123. This suggests that mitochondrial transmembrane potential is transiently elevated early during necrosis, at the time of cell and mitochondria swelling and prior to rupture of the plasma membrane [136].

A drop in mitochondrial transmembrane potential during apoptosis is revealed by a decrease in retention of Rh 123 [72, 74, 75, 137] as well as other mitochondrial potential probes such as carbocyanine dyes or the J-aggregate forming lipophilic cation JC-1 [71, 74, 138]. The latter fluorochrome appears to be a more specific probe of mitochondrial potential compared to Rh 123 or the carbocyanine dyes, and offers some other advantages. Uptake of JC-1 by mitochondria is reflected by a change in fluorescence emission from green to orange due to reversible formation of dye aggregates upon mitochondrial membrane polarization [71, 74, 138].

The mitochondrial changes appear to be ubiquitous for apoptosis, regardless of the mechanism of induction of apoptosis (e.g., engagement of Fas or DNA damage). They are also among the earliest occurrences of apoptosis. The loss of mitochondrial membrane integrity, regulated by the varying ratio of the pro- vs antiapoptotic members of the Bcl-2 family of proteins, may be directly and causally associated with the mechanism triggering the cascade of apoptotic effectors, proteases, and subsequently endonucleases. Analysis of the mitochondrial changes accompanying apoptosis, therefore, is expected to come into wide use in necrobiology.

It should be noted, however, that the uptake of the cationic probes by mitochondria depends not only on the mitochondrial transmembrane potential but also on their transport through the plasma membrane as well as their intracellular retention. Cells containing a very active glycoprotein P efflux pump (e.g., stem cells or multidrug resistant tumor cells) very efficiently pump out these fluorochromes thereby reducing their intracellular concentration. Consequently, the mitochondrial uptake of these probes is reduced. It is difficult, therefore, to assay apoptosis of cells overexpressing the efflux mechanisms by measuring the uptake of mitochondrial transmembrane potential probes.

4.3.2
Lysosomes

Other organelles whose function changes during cell death, and which can be analyzed by flow cytometry, are lysosomes. These organelles can be probed by lysosomo-trophic probes. The most frequently used probe is acridine orange (AO). Incubation of live cells in the presence of $1-2$ µg/ml of this metachromatic fluorochrome results in the uptake of this dye by lysosomes within cells which fluoresce red [139]. The uptake is the result of an active proton pump in lysosomes: the high proton concentration (low pH) causes AO, which enters the lysosome in an uncharged form, to become protonated and thus entrapped in the organelle. The mechanism responsible for a shift from green to red fluorescence of AO in lysosomes is unknown. It is likely that the dye is concentrated in lysosomes and forms aggregates. In such an aggregated form, AO is known to exhibit red fluorescence [140]. This assay is useful for cells that have numerous active lysosomes, such as monocytes, macrophages, etc. The uptake of AO by lysosomes is not significantly changed early during apoptosis. However, late apoptotic and necrotic cells show weak green and minimal red fluorescence at that low AO concentration. It should be stressed that the lysosomal uptake of AO is observed only in live, nonfixed cells. Regardless of the nature of the fixative (crosslinking or precipitating), fixation abolishes the ability of lysosomes to concentrate this dye.

4.4
Changes in Nuclear Chromatin

4.4.1
Altered DNA Stainability with EB, SYTO, and LDS-751

In contrast to Hoechst dyes or 7-AAD, the uptake of the DNA fluorochromes such as EB, certain SYTO dyes, or LDS-751 by unfixed, live cells is decreased at early stages of apoptosis [85, 141]. The decrease is observed at the time when the cell membrane permeability to the Hoechst dyes or 7-AAD is actually increased and DNA fragmentation is not yet extensive. Most likely, therefore, the observed decrease in DNA stainability early during apoptosis reflects changes in the nuclear chromatin, perhaps associated with its condensation, which restricts DNA accessibility to these fluorochromes. Decreased DNA stainability, conferred by histones and other nuclear proteins extractable with 0.1 mol/l HCl, has been observed in the case of chromatin condensation which occurs during erythroid differentiation of leukemic cells [142] or during spermatogenesis [143].

Interestingly, a transient rise in the stainability of DNA in chromatin with some intercalating fluorochromes has been observed at early stages of apoptosis [131]. The increase was significant enough to cause the appearance of a "pseudo-hyperdiploid" peak on the DNA content frequency histograms of these cell populations [133]. It is possible that the observed increase reflects the initial steps of DNA cleavage. Binding of intercalating molecules, which involves DNA unwinding and extension, is restricted for topological reasons in the case of closed circular- or closed loop-DNA conformation. DNA nicking by endonuclease(s) during apoptosis, therefore, by releasing the topological stress on the DNA helix, may initially enhance DNA stainability with the intercalating dyes. However, when DNA fragmentation progresses and becomes more extensive at later stages of apoptosis, the small MW DNA fragments may leak out from the cell, or be removed via shedding of apoptotic bodies. This event is then reflected by a loss of DNA stainability, as will be discussed later. Because of the transient nature of the increase in DNA stainability, it is difficult to accept this feature as a marker of apoptosis that can be used to identify apoptotic cells in routine practice.

4.4.2
Increased In Situ DNA Sensitivity to Denaturation

Free DNA in aqueous solution at physiological pH and ionic strength is in the double-stranded conformation. Upon treatment with heat, alkali, or acid the strands separate. This is known as DNA denaturation or helix-coil transition and is a result of destruction of the hydrogen bonding between the paired bases of the opposite strands. Because GC base pairs confer higher stability due to an additional hydrogen bond compared to AT, the intrinsic sensitivity of free DNA to denaturation depends exclusively on its GC:AT ratio.

DNA in situ in chromatin is additionally stabilized by interactions with histones, other nuclear proteins, and the nuclear matrix. Studies on the stability of DNA in situ, therefore, provide insight into chromatin structure, making its possible to discern double helix stabilizing interactions. Changes in chromatin as associated with cell cycle progression, differentiation, and apoptosis are reflected by altered DNA stability to denaturation.

The sensitivity of DNA in situ to denaturation can be measured based on the metachromatic property of the fluorochrome AO. Under proper conditions this dye can differentially stain double-stranded (ds) vs single-stranded (ss) nucleic acids [140, 142, 144]. Namely, when AO intercalates into ds DNA it emits green fluorescence. In contrast, the products of AO interaction with ss DNA fluorescered. In this method, the cells are briefly pre-fixed in formaldehyde followed by ethanol postfixation. Cell fixation abolishes staining of lysosomes, as described earlier. Following removal of RNA from the fixed cells by their preincubation with RNase, DNA is partially denatured in situ by short cell exposure to acid. The cells are then stained with AO at low pH to prevent DNA renaturation. It has been previously shown that the sensitivity of DNA in situ to denaturation is higher in condensed chromatin of mitotic cells compared to the noncondensed chromatin of interphase cells [140]. Apoptotic cells, like mitotic cells, have a larger fraction of DNA in the denatured form, and more intense red and reduced green fluorescence, compared to nonapoptotic (interphase) cells; the latter stain strongly green but have low red fluorescence [55, 56]. An increased degree of DNA denaturation (single-strandedness) in apoptotic cells can also be detected immunocytochemically, using an antibody reactive with ssDNA [145].

The method of identification or apoptotic cells based on sensitivity of DNA to undergo denaturation [27, 55, 56, 102, 142] may be uniquely applicable in situations where internucleosomal DNA degradation is not apparent [93–99, 146] and thus when other methods of apoptotic cell identification (e.g., the ones that rely on detection of DNA degradation) may fail (Fig. 6). They cannot, however, discriminate between mitotic and apoptotic cells. The method utilizing AO has been extensively applied to identify abnormal (apoptotic-like) sperm cells from individuals with reduced fertility and to study genotoxic effects of antitumor drugs and environmental poisons during spermatogenesis [57, 58].

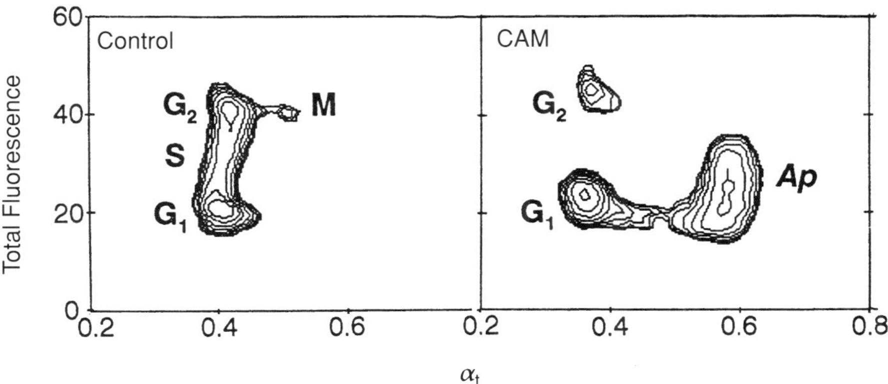

Fig. 6. Detection of apoptotic cells based on differences in sensitivity of DNA in situ to denaturation. HL-60 cells, untreated (Control) or treated with 0.15 μmol/l camptothecin (CAM) for 4 h were stained with AO as described [55]. Under the staining conditions used in this approach, the cellular green and red fluorescence intensities were proportional to the content of double- and single-stranded DNA, respectively, whereas total fluorescence corresponds to total DNA content (and thus correlates with the cell's position in the cycle) and α_t reflects the portion of denatured DNA. In the untreated culture, only mitotic cells have high red fluorescence and thus, a correspondingly high α_t ratio. CAM selectively triggered apoptosis of S phase cells, increasing their α_t ratio while the G$_1$ and G$_2$ cells were essentially unaffected

4.5
Cytometric Assays of DNA Fragmentation

Extensive DNA cleavage, being a characteristic feature of apoptosis, provided a basis for the development of two types of flow cytometric assays to identify apoptotic cells. One is based on extraction of low MW DNA prior to cell staining and identification of apoptotic cells as the cells with fractional DNA content [93, 147, 148]. This assay, introduced over 15 years ago [147] and modified afterwards [148] is still one of the most frequently used. The other, more recent assay, relies on fluorochrome labeling of DNA strand breaks in situ [104, 105, 114–116, 149–151].

4.5.1
Detection of Cells with Fractional DNA Content

Detection of apoptotic cells as the cells with fractional DNA content is a simple flow cytometric assay which requires cell permeabilization with detergents or prefixation with precipitating fixatives such as alcohols or acetone prior to their staining. Cell permeabilization or alcohol fixation does not fully preserve the degraded DNA within apoptotic cells: this fraction of DNA is extracted during subsequent cell rinsing and staining. As a consequence, apoptotic cells manifest reduced DNA content and therefore can be recognized following staining of cellular DNA with almost any nucleic acid specific fluorochrome, as cells with decreased DNA stainability ("sub-G_1" peak), located to the left of the peak representing G_1 cells on DNA content frequency histograms [147, 148].

The degree of DNA degradation varies depending on the stage of apoptosis, cell type, and often the nature of the apoptosis-inducing agent. The extractability of DNA during the staining procedure (and thus separation of apoptotic from live cells by this assay) also varies. It has been noted that addition of high molarity phosphate-citrate buffer to the rinsing solution enhances extraction of the degraded DNA [93].This approach can be used to control the extent of DNA extraction from apoptotic cells to the desired level to obtain the optimal separation of apoptotic cells by flow cytometry. Figure 2 illustrates an application of this approach to extract DNA from apoptotic cells for their subsequent identification by flow cytometry as well as for the electrophoretic analysis of DNA extracted from the very same cells studied by flow cytometry [42] discussed earlier.

Since measurement of DNA content provides information about the cell cycle position of the nonapoptotic cells, this approach can be applied to investigate the cell cycle specificity of apoptosis. Another advantage of this approach is its simplicity and applicability to any DNA fluorochrome or instrument. The combination of correlated DNA and RNA measurements, which allows one to identify G_0 cells, makes it possible to distinguish whether apoptosis is preferential to G_1 or G_0 cells [64].

It should be stressed, however, that the "sub G_1" peak can represent, in addition to apoptotic cells, mechanically damaged cells, cells with lower DNA content (e.g., in a sample containing cell populations with different DNA indices), or cells with different chromatin structure (e.g., cells undergoing erythroid differentiation) in which the accessibility of DNA to the fluorochrome is diminished. As will be discussed further, this is of special concern when unfixed cells are lysed in hypotonic solution, resulting in isolation of multiple nuclear fragments. Hence, the number of "sub G_1" particles in such a preparation represents the number of nuclear fragments and provides no information on the number of apoptotic cells. Furthermore, in addition to nuclear fragments, individual chromosomes released from mitotic cells, cell debris, and micronuclei may be mistakenly identified as apoptotic cells, especially when using a logarithmic scale of DNA content for data accumulation and analysis.

4.5.2
In Situ Labeling of DNA Strand Breaks

Endonucleolytic DNA cleavage results in the presence of extensive DNA cleavage in the chromatin of apoptotic cells. The 3' OH ends in DNA breaks can be detected by attaching to them biotin or digoxygenin conjugated nucleotides in a reaction catalyzed by exogenous TdT ("end-labeling", "tailing", "TUNEL") or DNA polymerase (nick translation) [104, 105, 113–116, 149–151].

Fluorochrome conjugated avidin or digoxygenin antibodies have often been used in the second step of the reaction to label DNA strand breaks. A simplified, single step procedure has also been proposed using fluorochromes directly conjugated to deoxynucleotides [151]. It is simpler compared to the indirect methods, but less sensitive.

A new method was recently introduced in which BrdUTP, incorporated into DNA strand breaks by TdT, is used as the marker of DNA strand breaks [149, 150] (Fig. 7). A similar approach has been used previously to probe the sensitivity of nuclear DNA to DNase I as a probe of nuclear structure [152]. The method based on BrdUTP incorporation [59] is simpler, more sensitive, and costs less compared with digoxygenin or biotin labeling. Commercial kits designed to label DNA strand breaks for identification of apoptotic cells are offered by ONCOR Inc. (Gaithersburg, MD, USA; two step assay using digoxygenin) and Phoenix Flow Systems (San Diego, CA, USA, two step, single step and BrdUTP labeling assays).

Cell fixation with a crosslinking fixative such as formaldehyde is essential for detection of DNA strand breaks [104, 105]. This is especially critical in the case of cells more advanced in apoptosis, containing extensively degraded DNA. Their DNA, unless crosslinked to intracellular proteins, can be washed out during the staining procedure, resulting in significant loss of both DNA content as well as the number of 3' OH termini which serve as primers for the TdT reaction.

The approach based on DNA strand break labeling in the assay employing TdT appears to be the most specific in terms of positive identification of apoptotic cells. Specifically, necrotic cells or cells with primary breaks induced by DNA damaging drugs or by ionizing radiation (up to the dose of 25 Gy of radiation) have an order of magnitude fewer DNA strand breaks than apoptotic cells [104]. Because the cellular DNA content of not only the nonapoptotic but also the apoptotic cells is measured, the method offers the unique possibility to analyze the cell cycle position, and/or DNA ploidy, of apoptotic cells directly [153].

It has been reported that comparative labeling of DNA strand breaks utilizing DNA polymerase vs TdT allows one to discriminate between apoptotic and necro-

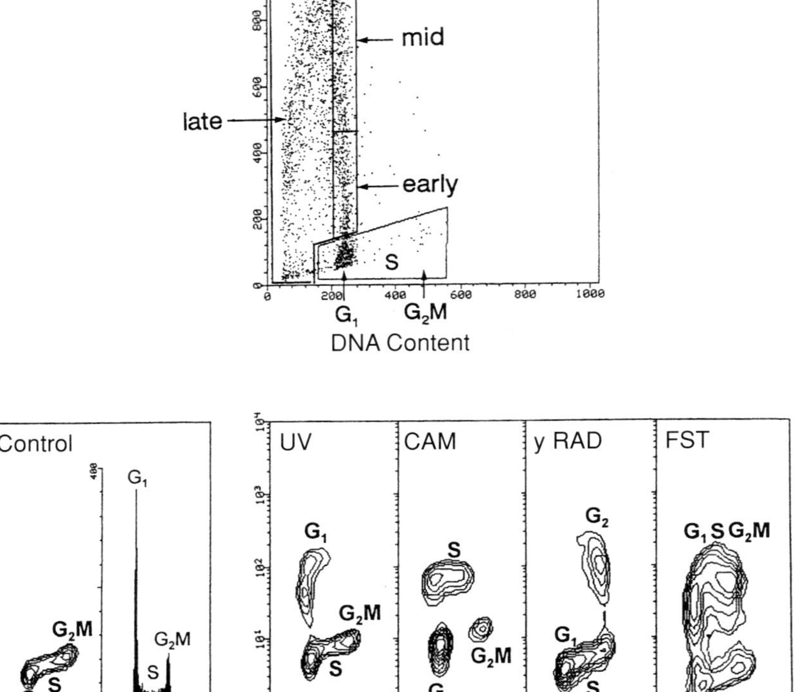

Fig. 7. DNA strand break labeling of human myeloid HL-60 leukemic cells and cells from a leukemia patient receiving taxol. HL-60 cells (*left*) were untreated (control) as well as treated with UV light, 0.15 µmol/l camptothecin (CAM), radiation (3 Gy) or DNA topoisomerase II inhibitor fostriecin (FST), and measured 4 h after the cells' exposure to drugs or irradiation. DNA strand breaks were labeled with digoxygenin-conjugated dUTP followed by FITC-digoxygenin MoAb. The cells were counterstained with PI. Note that UV light, CAM or radiation selectively induce apoptosis of G_1, S or G_2/M cells, respectively, while FST is less selective in terms of the cell cycle phase specificity. An example of apoptosis of leukemic cells (*top*) in the peripheral blood of a patient with acute leukemia undergoing chemotherapy with taxol. The data show the presence of cells with DNA strand breaks in relation to their DNA content. Note the heterogeneity of the apoptotic cells with respect to the number of DNA strand breaks per cell which most likely reflects the variability in the duration of apoptosis. After reaching the maximum number of strand breaks, cells lost DNA perhaps by shedding apoptotic bodies. With a loss of DNA, fewer primers (3' OH ends of DNA strand breaks) remain to be labeled with dUTP. (Adapted with permission from [1])

tic cells [35]. The difference in intensity of labeling of apoptotic and necrotic cells in these assays, however, was inadequate to separate these populations fully [104].

Because the procedure of DNA strand break labeling is rather complex and involves many reagents, negative results may not necessarily mean the absence of DNA strand breaks but may be the result of some methodological problem, such as the loss of TdT activity, degradation of triphosphonucleotides, etc. It is always necessary, therefore, to include a positive and negative control. An excellent control consists of HL-60 cells treated (during exponential growth) for 3–4 h with 0.2 µmol/l of the DNA topoisomerase I inhibitor camptothecin (CAM). Because CAM induces apoptosis selectively during S phase, the populations of G_1 and G_2/M cells serve as negative populations (background), while the S phase cells in the same sample, serve as the positive control Fig. 7).

The detection of apoptotic cells based on the presence of DNA strand breaks can be combined with analysis of DNA replication [61]. The advantage of this approach is that it offers the possibility, in a single measurement, to identify the cells which incorporated halogenated DNA precursors and the cells undergoing apoptosis in relation to their DNA content.

Bivariate analysis of DNA strand breaks vs cellular DNA content is the method of choice to study the cell cycle specificity of apoptosis [153]. Because the cells are briefly pre-fixed in a crosslinking fixative, the loss of DNA from early apoptotic cells is minimal. It is possible, therefore, to analyze the cell cycle distribution of both apoptotic, and non-apoptotic cell populations. Such analysis can discriminate between homo-phase or homo-cycle apoptosis vs post-mitotic apoptosis, as discussed earlier. This approach can also determine whether apoptosis is selective to cells of a particular DNA ploidy, e.g., whether in aneuploid tumors such tumor cells are more sensitive to apoptosis relative to normal cells, or in multiploid tumors which tumor cell population is more sensitive to the treatment.

As noted above, it is possible, using a recently introduced methodology, to identify apoptotic cells and cells replicating DNA in the same sample [150]. The method is based on prior incubation of cells with BrdU which is incorporated into DNA. Subsequently, the DNA strand breaks in apoptotic cells are labeled with a fluorochrome of a particular color. The cells are then subjected to UV light which induces photolysis of DNA containing the incorporated BrdU. The photolysis-generated DNA strand breaks are, in turn, labeled with a fluorochrome of another color, while DNA is counterstained with still another color fluorochrome [150]. In this way, in a single sample, one can identify apoptotic and BrdU incorporating cells, and classify them according to their respective phases of the cell cycle [150]. This method may be particularly useful in analysis of drug effects, when one has to distinguish between their cytostatic action, in terms of cell arrest in a particular phase of the cycle vs cytotoxicity, and in relation to cell cycle position.

5
Which Method to Choose?

The choice of a particular method for identification of dead cells depends on the cell type, the nature of the inducer of cell death, the mode and stage of cell death, the specific information that is desired (e. g., specificity of apoptosis with respect to cell cycle phase or DNA ploidy), and the technical possibilities and limitations such as the need for sample storage or transportation, the available instrumentation, choice of fluorochrome, etc.

Internucleosomal DNA fragmentation detected by agarose gel electrophoresis ("DNA laddering") is considered to be a very specific marker of apoptosis. The amount of the degraded DNA, especially within the range of mono- and oligonucleosomal DNA fragments, increases with progression of apoptosis. The intensity of DNA bands on gels, therefore, reflects both the number of apoptotic cells and the degree of DNA cleavage per individual cell. This approach, therefore, cannot be quantitative and should be used primarily to confirm the apoptotic mode of cell death, rather that to evaluate the frequency of apoptosis. Other methods of analysis of low MW DNA extracted from the cells in bulk, such as those based on immunochemical detection of nucleosomes or colorimetric quantitation of the "soluble" DNA fraction, cannot provide information about the frequency of apoptotic cells in cell populations.

It should be stressed that, because in many cell types DNA degradation during apoptosis does not progress to internucleosomal segments but stops at 300–50 kb fragments, the absence of the former segments (no "DNA laddering") is not evidence of the lack of apoptosis. The large DNA fragments, however, which can be detected by FIGE, do provide evidence of the apoptotic mode of cell death.

Gel electrophoresis of DNA from individual cells ("comets" assay) is an attractive methodology which allows one to evaluate the degree of DNA degradation in individual cells [154]. The "comets" can be evaluated using an image analyzing microscope, to obtain information about the proportion of degraded DNA and, from the distance between the comet's head and tail, about the approximate MW of the degraded DNA. The procedure, however, is very tedious and, therefore, not well suited to score the frequency of apoptotic cells in cell populations.

The methods based on analysis of plasma membrane integrity (exclusion of charged fluorochromes such as PI or hydrolysis of FDA), although simple and inexpensive, fail to identify apoptotic cells, especially at early stages of apoptosis. They can be used to identify necrotic cells, cells damaged mechanically, or cells advanced in apoptosis. Their major use is in enumeration of live cells in cultures (e.g., to exclude dead cells during analysis of cell growth curves, as when using the trypan blue exclusion assay), or in discrimination of mechanically broken cells (e.g., following mechanical cell isolation from solid tumors). Careful analysis of the kinetics of cell stainability with these dyes, however, may in some cell systems distinguish apoptotic cells.

The major advantage of the methods of identification of apoptotic cells based on analysis of the uptake of Hoechst fluorochromes or 7-AAD is their simplicity and low cost. They can be combined with the analysis of light scatter as well as with immunophenotyping. Light scatter serves then as an additional parameter discriminating apoptotic cells. These methods, however, require fresh, unfixed cells, which limits their usefulness when either longer storage or transportation of the samples are required. Furthermore the kinetics of uptake of the

Hoechst dyes or 7-AAD varies depending on the cell type and, with some cells, it is difficult to find the optimum conditions for discrimination of apoptotic from nonapoptotic cells.

The method based on cell reactivity with conjugates of annexin V and fluorochromes appears to be very specific and to detect cells in the early stages of apoptosis. Another virtue of this approach is its simplicity. However, as with other methods based on probing the plasma membrane, this method also requires fresh cells which precludes their prolonged storage or transport prior to analysis. Furthermore, as noted above, methods relying on probing plasma membrane integrity, permeability, or composition are not capable of discriminating late apoptotic cells from necrotic or mechanically broken cells. At the late stage of apoptosis, when plasma membrane integrity is destroyed (which corresponds to the stage of "apoptotic necrosis" as defined by Majno and Joris [2]), apoptotic cells may stain identically to necrotic cells with many fluorochromes and also be permeable to the annexin V-fluorochrome conjugates.

The presence of a large number of DNA strand breaks is a very specific marker of apoptosis [104]. Their detection, by the assay employing exogenous TdT, has therefore, become commonplace in studies of cell death. Similar to DNA laddering, however, the utility of this technique as a marker of apoptosis is complicated when dealing with atypical apoptosis, characterized by the lack of internucleosomal DNA degradation. The extent of DNA degradation in such atypical apoptotic cells is not always adequate for their identification by DNA strand break labeling. The data of Chapman et al. [155], however, indicate that, at least in some instances, apoptotic cells can be distinguished by the DNA strand breaks assay even in the absence of internucleosomal DNA cleavage. One should keep in mind, however, that though very rare, false positive recognition of apoptosis by the methods that rely on DNA fragmentation may occur in situations where internucleosomal DNA cleavage accompanies necrosis – e. g., [95].

Identification of apoptosis is better assured if more than a single viability assay is used. Simultaneous assessment of plasma membrane integrity (e.g., exclusion of charged fluorochromes or hydrolysis of FDA), together with either membrane permeability (Hoechst 33342 uptake), mitochondrial transmembrane potential (Rh 123 uptake), reactivity with the fluoresceinated annexin V, DNA sensitivity to denaturation, or DNA cleavage assays, offer more certainty of identifying the mode of cell death than each of these methods alone.

The choice of a method may also be dictated by its cost and complexity. Among flow cytometric assays, the least expensive and most rapid are the methods of discrimination of apoptotic cells based on DNA content analysis. These methods, therefore, may be preferred for screening drug effects in vitro, when large numbers of samples have to be rapidly analyzed. In addition to the enumeration of apoptotic cells, the cell cycle specific effects (cell cycle phase selective apoptosis or arrest in the cycle) are revealed in these methods by examination of the cellular DNA content frequency histograms representing the nonapoptotic cell populations. The mode of cell death, however, should be confirmed by analysis of the presence of DNA strand breaks or cell morphology.

Flow cytometric assays provide rapid, quantitative, and objective assessment of cell death and can be used for enumeration of apoptotic or necrotic cells. It should be stressed, however, that, regardless of the particular assay that has been used, flow cytometric analysis should always be confirmed by the inspection of cells under the microscope. Morphological changes during apoptosis are unique and they should be the deciding factor when ambiguity arises regarding the mechanisms of cell death. Furthermore, apoptosis was originally defined based on the analysis of cell morphology [7]. Morphological criteria for the identification of apoptotic and necrotic cells, therefore, should always be considered in conjunction with flow cytometric analysis of cell death, and be overriding in doubtful situations.

6
Inappropriate Uses of the Methodology

Identification of the mode of cell death and quantitation of apoptosis is not always simple and straightforward. There are many potential pitfalls and traps in analysis of the mode of cell death or in quantitation of dead cells, whether apoptotic or necrotic. The features of apoptotic cells vary significantly, depending on the cell (tissue) type, the nature of the apoptotic trigger, and the stage of apoptosis. Methods based on the detection of particular features of apoptosis, therefore, may fail if the cells undergoing apoptosis do not exhibit these features. Furthermore, apoptosis is a very transient event of variable duration, ending in total cell disintegration. Regardless of the method used, therefore, the measurement represents a single snapshot from which it is impossible to estimate the kinetics of apoptosis in cell populations. No method exists to obtain a cumulative measure of apoptosis such as can be done with mitotic cells in stathmokinetic experiments in which mitosis may be arrested by spindle poisons allowing one to estimate the rate of cell entrance to mitosis, the so-called "cell birth rate". The most common problems associated with the identification and quantification of apoptotic cells are discussed below.

6.1
Erroneous Quantitation of Apoptosis Based on the Extent of DNA Degradation

The amount of DNA degraded during apoptosis can be estimated by a variety of methods, including direct quantitative colorimetric analysis of the "soluble" DNA, densitometry of "DNA ladders" on gels, immunochemical assessment of nucleosomes, etc. Some of these approaches are advertised as quantitative in that they provide information regarding the frequency of apoptosis in cell populations. This is not correct. The amount of degraded, low MW DNA from each cell varies depending on the stage of apoptosis: early during apoptosis only a small DNA fraction is degraded but when apoptosis is advanced nearly all DNA is fragmented. Thus, the amount of degraded DNA from a single cell may vary many-fold depending on the stage of apoptosis. The total content of low MW DNA extracted from the cell population, or the ratio of low- to high-MW fractions do not provide information about the proportion of apoptotic cells, even in relative terms, e.g., for comparison of cell populations.

6.2
Erroneous Classification of Apoptotic Bodies or Chromatin Fragments as Apoptotic Cells

As mentioned, one of the most commonly used methods of identification of apoptotic cells relies on cellular DNA content measurement by flow cytometry. Particles with a fractional DNA content are assumed to be apoptotic cells. This, however, holds true for cells which were prefixed with precipitating fixatives such as alcohols or acetone. However, this analysis is commonly performed on cells which were subjected to treatment with a detergent or hypotonic solution instead of fixation. Such treatment lyses the plasma membrane. Because the nucleus of an apoptotic cell is fragmented and contains numerous individual chromatin fragments, the percentage of objects with a fractional DNA content (represented by the"sub-G_1" peak) released from a lysed cell may significantly exceed the apoptotic index. Furthermore, chromosomes or micronuclei are released upon lysis of mitotic cells or cells containing micronuclei, respectively. Therefore, individual chromosomes, chromosome aggregates, micronuclei, etc., all of which have a fractional DNA content, may erroneously be identified as apoptotic cells.

A gentle permeabilization of the cell with a detergent, but in the presence of exogenous proteins such as serum or serum albumin, prevents lysis of the plasma membrane. The presence of 1% (w/w) albumin or 10% (v/w) serum protects cells from lysis induced by, for example, 0.1% Triton X-100 without affecting their permeabilization by detergent. In fact, this method is used for simultaneous analysis of DNA and RNA as well as for detection of apoptotic cells characterized by reduced DNA content [64]. Apoptotic or nonapoptotic cells suspended in saline containing a detergent and serum proteins, however, can be very fragile and pipetting, vortexing, or even shaking the tube containing the suspension, may cause their lysis and release of the cell constituents into solution.

A logarithmic scale (log amplifiers) is frequently used to measure and display cellular DNA content in the methods which employ detergents to quantitate apoptosis. A logarithmic scale allows one to measure and record events with 1% or even 0.1% of the DNA content of intact, nonapoptotic cells. Most of these objects cannot be apoptotic cells. In the case of cell lysis by detergents, as discussed above, these objects represent nuclear fragments, individual apoptotic bodies, chromosomes, chromosome aggregates or micronuclei.

To exclude such objects, it is advisable during fluorescence measurement to set the threshold of DNA detection at a constant, e.g., one-tenth of the fluorescence value of intact G_1 cells, thereby excluding all particles with a DNA content less than 10% of G_1 cells. This may result in a slight underestimate of the apoptotic index but the bias is constant and it introduces less error compared to counting all objects with a fractional DNA content, especially those objects with DNA values between 0.1% and 1.0% of intact G_1 cells. In the latter case cellular debris, single chromosomes from broken mitotic cells, chromosome clumps, contaminating bacteria, etc., all having very low DNA content, may be erroneously classified as apoptotic cells. Thus, application of a linear rather than a logarithmic scale provides better assurance that objects with a minimal DNA content are excluded from the analysis.

6.3
Erroneous Assumption that the Percentage of Apoptotic Cells Represents the Cell Death Rate

Apoptosis is of short and variable duration. The time-window during which apoptotic cells demonstrate the characteristic features that allow them to be recognizable varies depending on the method used, cell type, or nature of the inducer of apoptosis. Some inducers may slow down or accelerate the apoptotic process by affecting the rate of formation and shedding of apoptotic bodies, endonucleolysis, proteolysis, etc., thus altering the duration of the "time window" by which we identify the apoptotic cell. An observed two-fold increase in apoptotic index may either indicate that twice as many cells were dying by apoptosis, compared to control, that the same number of cells were dying but that the duration of apoptosis was prolonged twofold, or a mixture of both phenomena. Unfortunately, no method exists to obtain cumulative estimates of the rate of cell entrance to apoptosis as there is, for example, for mitosis, which can be arrested by microtubule poisons in a stathmokinetic experiment. In short, the percentage of apoptotic cells in a cell population estimated by a given method is neither a measure of the rate cells are dying by apoptosis, nor a quantitative marker of cell death.

To estimate the rate (kinetics) of cell death the absolute number (not the percentage) of live cells should be measured in the culture, together with the rate of cell proliferation. The latter may be obtained from the rate of cell entrance to mitosis (cell "birth rate") in a stathmokinetic experiment. The observed deficit in the actual number of live cells from the expected number of live cells estimated based on the rate of cell birth provides the cumulative measure of cell loss (death). Indirectly, the cell proliferation rate can be inferred from the percentage of cells incorporating BrdUrd or from the mitotic index, under the assumption that the treatment which induces apoptosis does not affect the duration of any particular phase of the cell cycle (generally a risky assumption) and that the growth fraction is 100% (though this rarely is the case in clinical specimens).

6.4
Erroneous Assumption that Apoptotic Cells Exhibit All Classical Features of Apoptosis

The lack of evidence of apoptosis, detected by a particular method, is not necessarily evidence of the lack of apoptosis. As already discussed, there are numerous examples in the literature where cells die by a process resembling apoptosis which lacks one or more typical apoptotic features. Most frequently it occurs when DNA cleavage stops after generating 50–300 kb fragments. Such cells contain relatively few in situ DNA strand breaks compared with classical apoptosis [94–98, 156, 157]. The method of identification of apoptosis based on detection of the missing feature (e.g., DNA laddering on gels) fails to identify atypical apoptosis in such a situation.

Application of more than one method, each based on a different principle (i.e., detecting a different cell feature) stands a better chance of detecting atypical apoptosis than any single method. One expects, for example, that if DNA in apoptotic cells was fragmented to 50 kb it would not be extractable and, as a result, such cells could not be identified as apoptotic either by the method based on analysis of DNA content or that based on "laddering" DNA during electrophoresis. However, the presence of in situ DNA strand breaks, even at a lower frequency than in the case of typical apoptosis, can be used as a marker in such cells [106, 116]. It is likely (although not proven in many cell systems) that such apoptotic cells may additionally be recognized based on their reduced F-actin stainability with FITC-phalloidin [67], by their reactivity with FITC-annexin V conjugate [86–89], or by the drop in mitochondrial transmembrane potential detected by the potential sensing fluorochromes [74, 75].

6.5
Problems with Distinction of Late Apoptotic vs Necrotic Cells

Late stage apoptotic cells resemble necrotic cells to such an extent that, to define them, the term "apoptotic necrosis" was proposed [2]. Since the ability of apoptotic cells to exclude charged cationic fluorochromes such as PI or trypan blue is lost at these later stages, the discrimination between late apoptosis and necrosis cannot be accomplished by methods utilizing these dyes. Furthermore, because phosphatidylserine may be accessible to the annexin V-fluorochrome conjugates in necrotic cells due to the rupture of the plasma membrane, it is expected that this assay cannot be used to distinguish apoptotic from necrotic cells. Other methods, therefore, should be used. In most cases the degree of DNA fragmentation is much less pronounced during necrosis than during apoptosis [104]. Therefore, extensive DNA fragmentation detected by DNA gel electrophoresis analysis of cellular DNA content, or in situ presence of numerous DNA strand breaks, may all serve as markers to distinguish late apoptotic from necrotic cells.

6.6
Selective Loss of Apoptotic Cells During Sample Preparation

Apoptotic cells detach from the surface of the culture flasks and float in the medium. The standard procedure of discarding the medium following trypsinization or EDTA-treatment of the attached cells may result in selective loss of apoptotic cells. Such loss may vary from flask to flask depending on handling of the culture, e.g., degree of mixing or shaking, efficiency in discarding the old medium, etc. Surprisingly, discarding detached cells prior to trypsinization is a common practice in studies of apoptosis, as is evident from many published papers. Needless to say, such an approach cannot be used for quantitative analysis of apoptosis. To estimate the apoptotic index in cultures of adherent cells, it is essential to add the floating cells to the trypsinized ones and measure them together.

Similarly, density gradient (e.g., using ficoll-hypaque solutions) separation of the cells may result in selective loss of dying and dead cells. The knowledge of any selective loss of dead cells in cell populations purified by such approaches is essential when one is studying apoptosis.

6.7
Use of Untested Commercial Kits

A plethora of commercial kits have recently become available to detect apoptosis and reagent companies are racing to introduce their "unique apoptosis detection kits". Some of these kits have solid experimental foundations and have been repeatedly tested on a variety of cell systems. Other kits, however, especially those advertised by vendors who do not fully explain the principle of detection of apoptosis on which the kit is based, and do not list its chemical composition, may not be universally applicable. Unfortunately, it is a common practice that some vendors have tested only a single cell line using a single agent to trigger apoptosis (generally, either a leukemic cell line treated with the Fas ligand, or HL-60 cells treated with a DNA topoisomerase inhibitor). The claim is often made, however, that their kit is generically applicable to detect apoptosis.

Before using any new kit it is advisable to confirm that at least 3–4 independent laboratories have already successfully used it on different cell types. Furthermore, it is good practice initially to use the new kit in parallel with a well established methodology in a few experiments. This would allow one, by comparison of the apoptotic indices, to estimate the time-window of detection of apoptosis by the new method and its sensitivity, compared to the one which is already established and accepted in the field.

6.8
Failure to Examine Cell Morphology

This is one of the most common deficiencies in studies designed to reveal the mode of cell death, which utilize either flow cytometry or rely on analysis of DNA fragmentation in bulk (gel electrophoresis, immunochemical detection of nucleosomes). Apoptosis was originally defined as a specific mode of cell death based on very characteristic changes in cell morphology [54]. Although individual features of apoptosis may serve as markers for detection and analysis of the proportion of apoptotic cells in the cell populations studied by flow cytometry or other quantitative methods, the mode of cell death should always be positively identified by inspection of cells by light or electron microscopy. Therefore, when quantitative analysis is done by flow cytometry, it is essential to confirm the mode of cell death based on morphological criteria. Furthermore, if there is any ambiguity regarding the mechanism of cell death, the morphological changes should be the deciding attribute in resolving the uncertainty.

It should be stressed that optimal preparations for light microscopy require cytospinning of live cells followed by their fixation and staining on slides. The cells are then flat and their morphology is easy to assess. On the other hand, when the cells are initially fixed and stained in suspension, then transferred to slides and analyzed under the microscope, their morphology is obscured by the unfavorable geometry: the cells are spherical and thick and require confocal microscopy to reveal details such as early signs of apoptotic chromatin condensation.

Differential staining of cellular DNA and protein with DAPI and sulforhodamine 101 of the cells on slides, which is very rapid and simple, gives a very good morphological resolution of apoptosis and necrosis (Fig. 1). Other DNA fluorochromes, such as PI, 7-AAD, or AO, can be used as well.

Acknowledgment. Supported by NCI Grant RO1 28704, "This Close" Foundation for Cancer Research, and the Chemotherapy Foundation.

7
References

1. Darzynkiewicz Z, Juan G, Li X, Gorczyca W, Murakami T, Traganos F (1997) Cytometry 27:1
2. Majno G, Joris I (1995) Am J Pathol 146:3
3. Arends MJ, Morris RG, Wyllie AH (1990) Am J Pathol 136:593
4. Cohen JJ (1993) Immunol Today 14:126
5. Compton MM (1992) Cancer Metast Rev 11:105
6. Cotter TG (1992) Seminars Immunol 4:399
7. Kerr JFR, Wyllie AH, Curie AR (1972) Br J Cancer 26:239
8. Vaux DL (1993) Proc Natl Acad Sci USA 90:786
9. Wyllie AH (1992) Cancer Metast Rev 11:95
10. Wyllie AH, Arends MJ, Morris RG, Walker SW, Evan G (1992) Seminars Immunol 4:389
11. Oltvai ZN, Korsmeyer SJ (1994) Cell 79:189
12. Rowan S, Fisher DE (1997) Leukemia 11:457
13. Nagata S (1997) Cell 88:355
14. Golstein P (1997) Science 275:1081
15. Kumar S (1997) Cell Death Diff 4:2
16. Hockenberry DM (1995) J Cell Sci 18:51
17. Lomo J, Smeland EB, Krajewski S, Reed JC, Blomhoff HK (1996) Cancer Res 56:40
18. Reed JC (1994) J Cell Biol 124:1
19. Fernandes-Alnembri T, Takahashi A, Armstrong R, Krebs J, Fritz L, Tomaselli KJ, Wang L, Yu Z, Croce CM, Salveson G, Earnshaw WC, Litwack G, Alnemri ES (1995) Cancer Res 55:6045
20. Lazebnik YA, Kaufmann SH, Desnoyers S, Poirier GG, Earnshaw WC (1994) Nature 371:346
21. Nicholson DW, All A, Thornberry NA, Vallancourt JP, Ding CK, Gallant M, Gareau Y, Griffin PR, Labelle M, Lazebnik YA, Munday NA, Raju SM, Smulson ME, Yamin T-T, Yu VL, Miller DK (1995) Nature 376:37
22. Alnemri ES, Livingston DJ, Nicholson DW, Salvesen G, Thomberry NA, Wong WW, Juan J (1996) Cell 87:171
23. Bruno S, Lassota P, Giaretti W, Darzynkiewicz Z (1992) Oncology Res 4:29
24. Gorczyca W, Bruno S, Darzynkiewicz RJ, Gong J, Darzynkiewicz Z (1992) Int J Oncol 1:639
25. Hara S, Halicka HD, Bruno S, Gong J, Traganos F, Darzynkiewicz Z (1996) Exp Cell Res 232:372
26. Weaver VM, Lach B, Walker PR, Sikorska M (1993) Biochem Cell Biol 71:488
27. Chinnaiyan AM, O'Rourke K, Yu G-L, Lyons RH, Gang M, Duan RD, Genz R, Ni J, Dixit VM (1996) Science 274:990
28. Kiston J, Rawen T, Jiang Y-P, Goeddel DV, Giles KM, Pun K-T, Grinham CJ, Brown R, Farrow SN (1996) Nature 384:372

29. Krippner A, Matsuno-Yagi A, Gottlieb R, Babior B (1996) J Biol Chem 271:21,629
30. Evan DL, Wyllie AH, Gilbert CS, Littlewood TD, Land H, Brooks M, Waters CM, Penn LZ, Hancock DC (1992) Cell 69:11
31. Clarke AR, Purdie CA, Harrison DJ, Morris RG, Bird CC, Hooper ML, Wyllie AH (1993) Nature 362:849
32. Paulovich AG, Toczyski DP, Hartwell LH. (1997) Cell 88:315
33. Ray CA, Black RA, Kronheim SR, Greenstreet TA, Sleath PR, Salvesen GS, Pickup DJ (1992) Cell 69:597
34. Williams GT, Smith CA (1993) Cell 74:777
35. Darzynkiewicz Z (1995) J Cell Biochem 58:151
36. Dive C, Hickmann JA (1991) Br J Cancer 64:192
37. Fisher DE (1994) Cell 78:539
38. Hannun YA (1997) Blood 89:1845
39. Hickmann, J.A. (1992) Cancer Metast Rev 11:121
40. Kerr JFR, Winterford CM, Harmon V (1994) Cancer 73:2013
41. Hercbergs A, Leith JT (1993) JNCI 85:1342
42. Isaacs JT (1993) Envir Health Perspectives 101:(Suppl. 5) 27
43. Laidlaw IJ, Potten CS (1992) J Pathol 167:25
44. Gorczyca W, Bigman K, Mittelman A, Ahmed T, Gong J, Melamed MR, Darzynkiewicz Z (1993) Leukemia 7:659
45. Li X, Gong J, Feldman E, Seiter K, Traganos F, Darzynkiewicz Z (1994) Leukemia and Lymphoma 13:65
46. Seiter K, Feldman EJ, Traganos F, Li X, Halicka HD, Darzynkiewicz Z, Lederman CA, Romero M-B, Ahmed T (1995) Leukemia 9:1961
47. Halicka HD, Seiter K, Feldman EJ, Traganos F, Mittelman A, Ahmed T, Darzynkiewicz Z (1997) Apoptosis 2:25
48. Linette GP, Korsmeyer SJ (1994) Curr Opin Cell Biol 6:809
49. Young JD-E, Leong LG, Liu C-C, Damiano A, Wall DA, Cohn ZA (1986) Cell 47:183
50. Echaniz P, de Juan MD, Cuadrado E (1995) Cytometry 19:164
51. Meyaard L, Otto SA, Jonker RR, Mijnster MJ, Keet, RPM Miedema F (1992) Science 257:217
52. Evenson DP, Darzynkiewicz Z, Melamed MR (1980) Science 210:1131
53. Gorczyca W, Traganos F, Jesionowska H, Darzynkiewicz Z (1993) Exp Cell Res 207:202
54. Gledhill BL, Darzynkiewicz Z, Ringertz NR (1971) J Reprod Fert 26:25
55. Hotz MA, Traganos F, Darzynkiewicz Z (1992) Exp Cell Res 201:184
56. Hotz MA, Gong J, Traganos, F, Darzynkiewicz Z (1994) Cytometry 15:237
57. Evenson DP, Jost LK, Baer RK (1993) Envir Molec Mutagen 21:144
58. Sailer BL, Jost LK, Erickson KR, Tajiran MA, Evenson DP (1995) Envir Molec Mutagenesis 25:23
59. Darzynkiewicz Z, Bruno S, Del Bino G, Gorczyca W, Hotz MA, Lassota P, Traganos F (1992) Cytometry 13:795
60. Telford WG, King LE, Fraker PJ (1994) J Immunol Meth 172:1
61. Darzynkiewicz Z, Li X, Gong J (1994) Meth Cell Biol 41:16
62. Darzynkiewicz Z, Li X (1996) Measurements of cell death by flow cytometry. In: Cotter TG, Martin SJ (eds) Techniques in apoptosis. A user's guide. Portland Press, London, p 71
63. Darzynkiewicz Z, Li X, Gong J, Hara S, Traganos F (1995) Analysis of cell death by flow cytometry. In: Studzinski GP (ed) Cell growth and apoptosis. A practical approach. University Press. Oxford, p 143
64. Darzynkiewicz Z, Li X, Gong J, Traganos F (1997) Methods for analysis of apoptosis by flow cytometry. In: Rose NR, de Macario EC, Folds JD, Lane HC Nakamura R (eds) Manual of clinical laboratory immunology, 5th edn. ASM Press, Washington, DC, p 334
65. Shapiro H (1995) Practical Flow Cytometry. Wiley-Liss, New York
66. Darzynkiewicz Z, Robinson JP, Crissman HA (eds) (1994) Flow cytometry, 2nd ednn, parts A and B. Academic Press, San Diego
67. Endersen PC, Prytz PS, Aarbakke J (1995) Cytometry 20:162

68. Yang E, Korsmeyer SJ (1996) Blood 88:388
69. Gajewski TF, Thompson CB (1996) Cell 87:589
70. Muchmore SW, Sattlet M, Liang H, Meadows RP, Harlan JE, Yoon HS, Nettelsheim D, Chang BS, Thompson CB, Wong SL, Ng SL, Fesik SW (1996) Nature 381:335
71. Cossarizza A, Kalashnikova G, Grassilli E, Chiappelli F, Salvioli S, Capri M, Barbieri D, Troiano L, Monti D, Franceschi C (1994) Exp Cell Res 214:323
72. Lizard G, Fournel S, Genestier L, Dgedin N, Chaput C, Flacher M, Mutin M, Panaye G, Revillard J-P (1995) Cytometry 21:275
73. Castedo M, Hirsch T, Susi SA, Zamzani N, Marchetti P, Macho A, Kroemer G (1996) J Immunol 157:512
74. Cossarizza A, Mussini C, Mongiardo N, Borghi V, Sabbatini A, De Rienzo B, Franseschi C (1997) AIDS 11:19
75. Petit PX, LeCoeur H, Zorn E, Dauguet C, Mignotte B, Gougeon ML (1995) J Cell Biol 130:157
76. Hedley DW, McCulloch EA (1996) Leukemia 10:1143
77. Yang J, Liu, X, Bhalla K, Kim CN. Ibrado AM, Cai J, PengT-I, Jones DP, Wang X (1997) Science 275:1129
78. Kluck RM, Bossy-Wetzel E, Green DR, Newmeyer DD (1997) Science 275:1132
79. McConkey DJ, Nicotera P, Hartzell P, Bellomo G, Wyllie AH, Orrenius S (1989) Arch Biochem Biophys 269:365
80. Zamzani N. Susin SA, Marchetti P, Hirsch T, Gomez-Monterey L, Castedo M, Kroemer G (1996) J Exp Med 183:1533
81. Chinnaiyan AM, O'Rourke K, Lane BR, Dixit VM (1997) Science 175:1122
82. Wu D, Wallen H, Nunez G (1997) Science 175:1126
83. Ormerod MG, Sun X-M, Snowden RT, Davies R, Fearhead H, Cohen GM (1993) Cytometry 14:595
84. Schmid I, Krall WJ, Uttenbogaart, Braun J, Giorgi JV (1992) Cytometry 13:204
85. Frey T (1995) Cytometry 21:265
86. Fadok VA, Voelker DR, Cammpbell PA, Cohen JJ, Bratton DL, Henson PM (1992) J Immunol 148:2207
87. Koopman G, Reutelingsperger CPM, Kuijten GAM, Keehnen RMJ, Pals ST, van Oers MHJ (1994) Blood 84:1415
88. van Engeland M, Ramaekers FCS, Schutte B, Reutelingsperger CPM (1996) Cytometry 24:131
89. Piacentini M, Fesus I, Farrace MG, Ghibelli L, Piredda L, Meline G (1995) Eur J Cell Biol 54:246
90. Enari M, Talanian RV, Wong WW, Nagata S (1996) Nature 380:723
91. Gorczyca W, Ardelt B, Darzynkiewicz Z (1993) Int J Oncol 3:627
92. Oberhammer F, Wilson JM, Dive C, Morris ID, Hickman JA, Wakeling AE, Walker PR, Sikorska M (1993) The EMBO J 12:3679
93. Gong J, Traganos F, Darzynkiewicz Z (1994) Anal Biochem 218:314
94. Cohen GM, Su X-M, Snowden RT, Dinsdale D, Skilleter DN (1992) Biochem J 286:331
95. Collins RJ, Harmon BV, Gobe GC Kerr JFR (1992) Int J Radiat Biol 61:451
96. Ormerod MG, O'Neill CF, Robertson D, Harrap KR (1994) Exp Cell Res 211:231
97. Zakeri ZF, Quaglino D, Latham T, Lockshin RA (1993) FASEB J 7:470
98. Zamai L, Falcieri E, Marhefka G, Vitale M (1996) Cytometry 23:303
99. Akagi Y, Ito K, Sawada S (1995) Int J Radiat Biol 64:47
100. Catchpoole DR, Stewart BW (1993) Cancer Res 53:4287
101. Fukuda K, Kojiro M, Chiu J-F (1993) Am J Pathol 142:935
102. Hara S, Halicka HD, Bruno S, Gong J, Traganos F, Darzynkiewicz Z (1996) Exp Cell Res 223:327
103. Gong J, Traganos F, Darzynkiewicz Z (1995) Cell Growth and Diff 6:1485
104. Gorczyca W, Bruno S, Darzynkiewicz R, Gong J, Darzynkiewicz Z (1992) Int J Oncol 1:639
105. Gorczyca W, Gong J, Darzynkiewicz Z (1993) Cancer Res 52:1945
106. Del Bino G, Skierski J, Darzynkiewicz Z (1991) Exp Cell Res 195:485

107. Chang WP, Little JB (1991) Int J Radiol 60:483
108. Shinohara K, Nakano H (1993) Rad Res 135:197
109. Del Bino, G. Skierski JS, Darzynkiewicz Z (1990) Cancer Res 50:5746
110. Matsubara K, Kubota M, Kuwakado K, Hirota H, Wakazono Y, Okuda A, Bessho R, Lin YW, Adachi A, Akiyama Y (1994) Exp Cell Res 213:412
111. Gorman A, McCarthy J, Finucane D, Reville W, Cotter TG (1996) Morphological assessment of apoptosis. In: Cotter TG, Martin SJ (eds) Techniques in apoptosis. A user's guide. Portland Press, London. p 1
112. Ramachandra S, Studzinski GP (1995) Morphological and biochemical criteria of apoptosis. In :Studzinski GP (ed) Cell growth and apoptosis. A practical approach. Oxford University Press, Oxford p 119
113. Gold R, Schmied M, Giegerich G, Breitschopf H, Hartung HP, Toyka KV, Lassman H (1994) Lab Invest 71:219
114. Gold R, Schmied M, Rothe G, Zischler H, Breitschopf H, Wekerle H, Lassman H (1993) J Histochem Cytochem 41:1023
115. Gavrieli Y, Sherman Y, Ben-Sasson A (1992) J Cell Biol 119:493
116. Wijsman JH, Jonker RR, Keijzer R, van de Velde CJH, Cornelisse CJ, van Dierendonck JH (1993) J Histochem Cytochem 41:7
117. Salzman GC, Singham SB, Johnston RG, Bohren CF (1990) Light scattering and cytometry. In: Melamed MR, Lindmo T, Mendelsohn ML (eds) Flow cytometry and sorting. Wiley-Liss, New York. p 81
118. Swat W, Ignatowicz L, Kisielow P (1981) J Immunol Meth 137:79
119. Ormerod MG, Cheetham FPM, Sun X-M (1995) Cytometry 21:300
120. Belloc F, Dumain P, Boisseau MR, Jalloustre C, Reiffers J, Bernard P, Lacombe F (1994) Cytometry 17:59
121. Jacobs DP, Pipho C (1983) J Immunol Meth 62:101
122. Lyons AB, Samuel K, Sanderson A, Maddy AH (1992) Cytometry 13:809
123. Philpott NJ, Turner AJC, Scopes J, Westby M, Marsh JCW, Gordon-Smith EC, Dalgeish AG, Gibson FM (1996) Blood 87:2244
124. Schmid I, Uttenbogaart CH, Giorgi JV (1994) Cytometry 15:12
125. Rotman B, Papermaster BW (1966) Proc Natl Acad Sci USA 55:134
126. Riedy MC, Muirhead KA, Jensen CP, Stewart CC (1991) Cytometry 12:133
127. Stewart CC, Stewart SJ (1994) Meth Cell Biol 41:39
128. Ormerod MG, Collins MKL, Rodriguez-Tarduchy G, Robertson D (1992) J Immunol Meth 153:57
129. Chrest FJ, Buchholtz MA, Kim YH, Kwon T-K, Nordin AA (1993) Cytometry 14:883
130. Dive C, Gregory CD, Phipps DJ, Evans DL, Milner AE, Wyllie AH (1992) Biochim Biophys Acta 1133:275
131. Elstein KH,Thomas DJ, Zucker RM (1995) Cytometry 21:170
132. Elstein KH, Zucker RM (1994) Exp Cell Res 211:322
133. Ferlini C, Di Cesare S, Rainaldi G, Malorni W, Samoggia P, Biselli R, Fattorossi A (1996) Cytometry 24:106
134. Boersma AWM, Nooter K, Oostrum RG, Stoter G (1996) Cytometry 24:125
135. Johnson LV, Walsh ML, Chen LB (1980) Proc Natl Acad Sci USA 77:990
136. Darzynkiewicz Z, Traganos F, Staiano-Coico L, Kapuscinski J, Melamed MR (1982) Cancer Res 42:799
137. Shimizu S, Eguchi Y, Kamiike Y, Waguri W, Uchiyama Y, Matsuda H, Tsujimo Y (1996) Oncogene 13:21
138. Cossarizza A, Ceccarelli D, Masini A (1996) Exp Cell Res 222:84
139. Traganos F, Darzynkiewicz Z (1994) Meth Cell Biol 41:185
140. Darzynkiewicz Z, Kapuscinski J (1990) Acridine orange, a versatile probe of nucleic acids and other cell constituents. In: Melamed MR, Lindmo T, Mendelsohn ML (eds) Flow cytometry and sorting, 2nd edn. Wiley-Liss, New York, p 291
141. Poot M, Gibson LL Singer V L (1997) Cytometry 27:358

142. Darzynkiewicz Z, Traganos F, Kapuscinski J, Staiano-Coico L, Melamed MR (1984) Cytometry 5:355
143. Evenson D, Darzynkiewicz Z, Jost L, Janca F, Ballachey B (1986) Cytometry 7:45
144. Darzynkiewicz Z (1994) Meth Cell Biol 41:527
145. Frankfurt OS, Byrnes JJ, Seckinger D, Sugarbaker EV (1993) Onc Res 5:37
146. Tounecti O, Pron G, Belehradek J, Mir LM (1993) Cancer Res 53:5462
147. Umansky SR, Korol BR, Nelipovich PA (1981) Biochim Biophys Acta 655:281
148. Nicoletti I, Migliorati G, Pagliacci MC, Grignani F, Riccardi C (1991) J Immunol Meth 139:271
149. Li X, Darzynkiewicz Z (1995) Cell Prolif 28:571
150. Li X, Melamed MR, Darzynkiewicz Z (1996) Exp Cell Res 222:28
151. Li X, Traganos F, Melamed MR, Darzynkiewicz Z (1995) Cytometry 20:172
152. Szabo G Jr, Damjanovich S, Sumegi J, Klein G (1987) Exp Cell Res 169:158
153. Gorczyca W, Gong J, Ardelt B, Traganos F, Darzynkiewicz Z (1993) Cancer Res 53:3186
154. Olive PL, Frazer G, Banath JP (1993) Radiat Res 136:130
155. Chapman R, Chresta CM, Herberg AA, Beere HM, Heer S, Whetton AD, Hickman J, Dive C (1995) Cytometry 20:245
156. Ueda N, Walker PD, Hsu S-M, Shah SV (1995) Proc Natl Acad Sci USA 92:7202
157. Schwartz LM, Smith SW, Jones ME, Osborne LA (1993) Proc Natl Acad Sci USA 90:980

Received October 1997

The Bcl-2 Gene Family and Apoptosis

E. M. Bruckheimer[1] · S. H. Cho[1] · M. Sarkiss[1] · J. Herrmann[1] · T. J. McDonnell[2]

[1] Department of Molecular Pathology. The University of Texas M. D. Anderson Cancer Center, Houston, TX 77030
[2] Department of Molecular Pathology – Box 89, The University of Texas M. D. Anderson Cancer Center, 1515 Holcombe Blvd, Houston, Texas 77030, *E-mail: tim_mcdonnell @path.mda.uth.tmc.edu*

Apoptosis, or programmed cell death, is an essential process for normal embryonic development, maintaining homeostasis in adult tissues, and suppressing carcinogenesis. The bcl-2 protein, discovered in association with follicular lymphoma, plays a prominent role in controlling apoptosis and enhancing cell survival in response to diverse apoptotic stimuli. The evolutionarily conserved bcl-2 protein is now recognized as being a member of a family of related proteins which can be categorized as death agonists or death antagonists. Progress in defining the role of bcl-2 and its family members in regulating apoptosis is rapidly advancing. This review describes, in detail, current bcl-2 family members and the possible mechanisms of function which allow the bcl-2 family of proteins to either promote or suppress cell death.

Keywords: bcl-2, bcl-X, bax, bak, NFkB, nuclear import, protease, ion channels, cancer, apoptosis.

Advances in Biochemical Engineering/
Biotechnology, Vol. 62
Managing Editor: Th. Scheper
© Springer-Verlag Berlin Heidelberg 1998

List of Symbols and Abbreviations

BH	bcl-2 homology domain
DNA-PK	DNA-dependent protein kinase
ER	endoplasmic reticulum
ICE	interleukin 1β converting enzyme
Ig	immunoglobulin
KO	knockout
NPC	nuclear pore complex
PARP	poly ADP-dependent polymerase
PEST	proline-glutamate-serine-threonine
PKC	protein kinase C
PT	permeability transition
ROS	reactive oxygen species
SM	sphingomyelin
Smase	sphingomyelinase
snRNPs	small nuclear ribonuclear proteins
SV	Sindbis virus
TG	thapsigargin
TNF	tumor necrosis factor α
TPA	phorbol 12-myristate 13-acetate
UTR	untranslated region

1
Introduction

Throughout development, the process of cell death has been universally accepted as a naturally occurring event necessary for metamorphosis and normal embryologic development. Elimination of interdigital tissues and the development of limbs and organs requires the destruction of large numbers of cells. This process needs to occur in a regulated fashion [1]. It has been demonstrated that the "programming" of cell death that takes place in embryonic development and that this programming is, at least to some extent, mediated by tissue diffusible molecules [1]. Studies of amphibian and invertebrate metamorphosis also suggests that hormones may be involved in the regulation of developmental cell death. Initially, two types of cell death were recognized from studies of developmental systems. One type occurs as a normal function of the developing organism and has been termed "natural" cell death [2]. The other mode of cell death usually results from toxic stress or tissue damage and has been referred to as "accidental" cell death [2]. Shrinkage necrosis is a type of natural cell death during which cells undergo nuclear and cytoplasmic condensation and release membrane-bound vesicles. The resulting cell- and membrane-bound vesicles are then phagocytosed by neighboring cells or macrophages. Shrinkage necrosis does not elicit an immune response and the integrity of the intracellular organelles remains intact. This has also been observed in single cells amidst populations of normal cells and can occur as a result of multiple physiological and non-physiological stimuli. Coagulative necrosis is a type of accidental cell death characterized by cellular and organelle swelling and subsequent rupture. The intracellular contents of the cell are released into the extracellular space, thereby triggering an immune response and inflammation. Coagulative necrosis affects groups of cells often resulting from injury or cytotoxic treatments. A detailed description of the ultrastructural features of coagulative and shrinkage necrosis is presented by Wyllie et al. [2].

Although these two forms of cell death had been recognized for some time, it was John Kerr, Andrew Wyllie, and Sir Alistair Currie who, in 1972, defined apoptosis as a type of cell death which complements mitosis in the regulation of cell populations [3]. Apoptosis can occur as a result of both physiologic and pathologic conditions and is believed to be, in many developmental contexts, a programmed event. The sequence of events beginning with nuclear and cytoplasmic condensation and ending with the release and phagocytosis of apoptotic bodies was often overlooked in many systems because it was believed to represent autophagocytosis. Their paper described a key role for apoptosis in the normal turnover of cells within normal adult tissues and in embryonic development as well as demonstrating apoptosis in cancer cells [3]. This landmark work defined a new field of scientific inquiry where understanding the mechanisms of cell death became as relevant as understanding the mechanisms of cell division.

Much has been accomplished in the field of cell death research since apoptosis was defined in 1972. A major advance in understanding the molecular regulation of cell death originated with the discovery of the bcl-2 proto-oncogene from the t(14;18) chromosomal translocation breakpoint in follicular lymphoma [4–8]. Subsequent to its discovery, it has been shown that bcl-2 acts to suppress cell death triggered by a variety of stimuli which will be discussed in detail. Also, it is now apparent that there is a family of bcl-2 cell death regulatory proteins which share in common structural and sequence homologies. These different family members have been shown to either possess similar functions to bcl-2 or counteract bcl-2 function and promote cell death.

Apoptosis is a highly conserved process throughout evolution. Studies with the nematode *Caenorhabditis elegans* have convincingly demonstrated this conservation. During *C. elegans* development, 131 of the initial 1090 somatic cells undergo programmed cell death. The process of cell death has been investigated through mutational analysis and several key genes and gene products have been discovered. In 1986, Ellis and Horvitz showed that ced-3 and ced-4 are required for all the somatic cell deaths occurring in *C. elegans* [9]. Through sequence homology, ced-3 has been shown to be homologous to the mammalian interleukin 1 β converting enzyme (ICE) family of cysteine proteases, or caspases [10] which are also involved in the mediation of apoptosis [11, 12]. Ced-4 has recently been demonstrated to link ced-3 to ced-9 in the apoptotic cascade [13]. Additionally, both ced-3 and ced-4 have been shown to induce cell death when expressed in mammalian cell lines [13, 14]. While ced-3 and ced-4 are necessary for cell death, it was shown that ced-9 is necessary for cell death suppression [15]. Ced-9 shares sequence homology to bcl-2 and human bcl-2 can, in large part, substitute for ced-9 function in *C. elegans* [15 – 17].

In mammalian development, bcl-2 and bcl-2 family members have been shown to play a role in morphogenesis and normal development. During murine fetal development, bcl-2 is expressed in tissues derived from all three germ layers; however, as the fetus matures, bcl-2 expression becomes restricted [18]. For example, bcl-2 is widely expressed throughout the undifferentiated intestinal epithelium in the fetus but expression is restricted to the progenitor cells when the villi mature [18]. Likewise, the expression patterns of bcl-2 observed during the murine retinal and renal development are in the undifferentiated tissues and the differentiated tissues have lower bcl-2 expression levels [18]. Similar observations were seen in human fetal tissues in that bcl-2 was expressed in a wide variety of tissue types and expression became restricted as the fetus matured [19, 20]. Bcl-2 was detected in the human fetal thymus, hematopoietic cells, endocrine glands, and hormonally regulated tissues. An example of differential expression of bcl-2 family members occurs during neuronal differentiation. Bcl-X_L and bcl-2 are both expressed in neurons of the developing human fetus; however, bcl-X_L expression persists throughout fetal development and into adulthood whereas bcl-2 expression diminishes between weeks 20 – 39 of gestation [21]. Although bcl-2 protein is widely expressed in embryonic tissues [18, 22], absence of bcl-2 protein in bcl-2 null mice does not interfere with normal prenatal development [23]. However, postnatally, these mice display growth retardation, smaller ears, polycystic kidneys, and most die within several months due to kidney failure. In the bcl-2 deficient mice, which eventually become ill, the thymus and spleen are atrophic due to massive lymphocyte apoptosis. Also, bcl-2 null thymocytes are more likely to undergo apoptosis following γ-irradiation or treatment with dexamethasone [24, 25].

The tissue distribution of bcl-2 expression also suggests that bcl-2 plays a role in survival in various cell types [26]. Immunohistochemistry reveals that bcl-2 is expressed in cells that regenerate such as the stem cells or in cells that are long lived. In the lymphatic system, bcl-2 is strongly expressed in the thymic medulla where the T-cells which have survived negative and positive selection reside, and in the areas of lymph nodes associated with maintenance of plasma cells and memory B-cells [26, 27]. In non-hematopoietic tissues, bcl-2 is restricted to cells that undergo self renewal such as the basal layer of the skin, the crypt cells of the small and large intestine, and in long lived cells such as the neurons. Bcl-2 is also expressed in tissues such as breast duct epithelium and prostate epithelium which undergo hyperproliferation or involution at the influence of the hormone or growth factors [26, 28].

The bcl-2 family continues to expand with the discovery of new members. This review will focus on the bcl-2 family of cell death regulators summarized in Table 1.

Table 1. Human bcl-2 family members

Family Member	Gene Size (Kb)	mRNA Size (Kb)	Amino Acid Residues	Protein Size (kD)	Chromosome Localization	Function
Bcl-2	230	6.5	239	25	18q21	Anti-Apoptotic
Bcl-x$_L$	ND	2.7	233	31	ND	Anti-Apoptotic
Bcl-w	22	3.7	193	22	ND	Anti-Apoptotic
Mcl-1	ND	3.8	350	37.3	1q21	Anti-Apoptotic
A1	ND	1.4[a]	172	20[b]	ND	Anti-apoptotic
Bfl-1	ND	0.6[a]	175	20[c]	15q24.3	Anti-Apoptotic
Bax α	4.5	1.0	192	21	19q13.3 – 13.4	Pro-Apoptotic
Bax β		1.5	218	24		ND
Bax γ		1.5	41[d]	4.5[b]		ND
Bax δ		ND	143[d]	?		ND
Bcl-x$_S$	ND	1.0	170	19	ND	Pro-Apoptotic
Bak 1	6	2.4	211	23	6	Pro-Apoptotic
Bak 2				20		
Bak 3				11		
Bad	ND	1.1[a]	204	22	ND	Pro-Apoptotic
Bid	ND	1.1[a]	195	23	ND	Pro-Apoptotic
Bik	ND	1.0	160	18	ND	Pro-Apoptotic
GRS	ND	0.8	175	?	15q24 – 25	ND
Harakiri	ND	0.7	91	16[e]	ND	Pro-Apoptotic

ND not determined.
[a] cDNA, [b] predicted size of protein, [c] size of Ha-Tagged protein, [d] predicted amino acid length, [e] size of the Flagged protein.

2
Genetic Determinants of Cancer and the Bcl-2 Oncogene

It is commonly accepted that tumorigenesis is a multistep process which may involve chromosomal abnormalities and the deregulated expression of proto-oncogenes [29]. This is particularly evident in hematolymphoid neoplasms. Characterization of the t(9;22) and t(8;14) translocations in chronic myelogenous leukemia [30, 31] and Burkitt's lymphoma [32, 33], respectively, provided a paradigm for the deregulation of proto-oncogenes during multistep carcinogenesis. The t(9;22) results in the formation of a bcr-abl fusion gene and chimeric protein [34] while the t(8;14) results in the inappropriate expression of c-myc [35–37]. Both of these molecular events result in augmented cellular proliferation [38].

Bcl-2 was discovered as a novel transcriptional element by its association with the t(14;18) reciprocal chromosomal translocation commonly found in follicular lymphoma [4, 5, 39]. Bcl-2 was shown to be a unique oncogene in that its deregulation did not result in an increase in cell proliferation, but rather enhancement of cell survival [40–42]. Thus, bcl-2 represents a new class of oncogene, one that enables neoplastic growth by suppressing cell death [43].

The bcl-2 gene, comprised of three exons, spans approximately 230 Kb. The open reading frame is in exon 2 and 3, and encodes a 25 kD integral membrane protein [44, 45]. Exon 1 possesses stop codons in all open reading frames. The intron between exon 1 and 2 is 220 bp and is alternatively spliced in both the normal bcl-2 mRNA and bcl-2-Ig fusion mRNAs. The second intron is in excess of 200 Kb. The t(14;18) major break point region is in the 3' untranslated region (UTR) of exon 3. Thus the bcl-2 open reading frame is not disrupted by the t(14;18), and generates an otherwise normal bcl-2 gene product in inappropriate amounts [44]. The initiation codon is in exon 2 and upstream of it are the promoter sequences TATAA (–88) and CCAAT (-106), and the initiation sites at –31 and –58. Also present is the sequence ATGCAAAGCA (–119), similar to an SV40 enhancer, and an upstream enhancer of Ig region. Upstream of exon I, multiple transcription initiation sites exist. This region is very GC rich and possesses seven Sp1 binding sites. S1 protection assays indicate that the upstream exon 1 initiation sites are used most frequently [44]. The message can be alternatively spliced to give two transcripts, bcl-2 α and the truncated bcl-2 β that lacks the C-terminus region [8].

Bcl-2 possesses a very hydrophobic stretch of 23 amino acids at the C-terminus which serve as a transmembrane domain [41]. Bcl-2 protein localizes to the nucleus, rough ER, and mitochondria. In mitochondria, the protein is localized to the contact zone of the inner and outer membranes of the mitochondrial membrane where the transport of materials from the cytosol into the mitochondrial matrix occurs [41, 46].

Bcl-2 is normally expressed in pro- and mature B cells, but is downregulated in pre and immature B lymphocytes [47]. This differential expression points to the survival role of bcl-2 in B lymphocyte development. High levels of bcl-2 are needed to ensure the survival of pro-B cells and mature B cells in order to maintain a population of functional lymphocytes. But low levels of bcl-2 are necessary for cells which do not express functional surface Ig or are self-reactive to undergo apoptosis. Also in T-cells, bcl-2 is expressed at low levels in double positive thymocytes undergoing negative and positive selection, and at high levels in mature single positive T-cells which have survived the selection [48]. Thus, bcl-2 seems to have an important role in lymphocyte development.

To study the effects of deregulated expression of bcl-2, transgenic mice were created using a bcl-2-Ig mini-gene mimicking the fusion gene resulting from the t(14;18) translocation in follicular lymphomas [42]. These animals initially developed a polyclonal expansion of mature B-cells bearing surface immunoglobulins M and D. When these resting cells were placed in vitro, they survived significantly longer than the lymphocytes derived from control littermate mice. Thus, the B-cell hyperplasia observed in the transgenic mice is due to the extended life span of the cell and not secondary to enhanced proliferation. Together, these observations support the contention that deregulation of bcl-2 is a primary pathogenic event during lymphomagenesis but malignant transformation per se is dependent on the occurrence of additional genetic events. By abnormally extending cellular life span, bcl-2 appears to render the cell more susceptible to acquiring additional genetic lesions that are capable of complementation with bcl-2 and resulting in malignant conversion [49].

Indeed, the bcl-2-Ig transgenic mice eventually developed malignant lymphomas after a long latency [50]. The lymphomas were most frequently clonal high grade B cell immunoblastic lymphomas. Approximately 50% of the immunoblastic lymphomas exhibit c-myc gene rearrangements as a consequence of t(12;15) chromosomal translocations. The t(12;15) resulted in a head-to-head joining of the c-myc gene with the immunoglobulin heavy chain locus and therefore recreated the molecular features of the t(8;14) translocation which is the cytogenetic hallmark of human Burkitt's lymphoma [51]. This transgenic animal model demonstrates that deregulation of the bcl-2 gene causes initially a polyclonal expansion of mature B-cells which can progress to an aggressive monoclonal malignancy with an acquisition of additional gene deregulation, thus confirming the multistep nature of carcinogenesis [52]. In humans also, follicular lymphoma can progress to a high grade lymphoma following the acquisition of t(8;14) translocations and c-myc gene deregulation, albeit this appears to be an uncommon event [53]. The ultimate test of c-myc and bcl-2 complementation during lymphomagenesis was the dramatically shortened latency to lymphoma development observed in Eμ-myc x bcl-2-Ig double transgenic mice [51]. The basis of c-myc and bcl-2 complementation appears to be the inhibition of c-myc mediated apoptosis without substantially altering the enhanced rate of cell division associated with deregulated c-myc [51, 54].

It has also been demonstrated that bcl-2 plays a role in the suppression of p53 mediated cell death. Splenic mononuclear cells obtained from bcl-2-Ig mice, which possess wild type p53, displayed rates of apoptosis comparable to cells obtained from p53 knockout mice following γ-irradiation [55]. Together, these results and the results of others utilizing transformed cell lines indicate that bcl-2 is capable of blocking p53 mediated cell death induction [55–57]. Mutations in the conserved domains of p53 were uncommon in the lymphomas arising in the bcl-2-Ig transgenic mice, suggesting that there is no selective advantage of acquiring p53 mutations when bcl-2 is overexpressed [55]. Additionally, the bcl-2-Ig transgenic and p53 knockout murine models were further utilized to determine the extent of genetic complementation between p53 and bcl-2. In p53 KO/bcl-2 hybrid mice, tumor latency and incidence were unchanged when compared to individual parental strains of mice [55]. Many human tumors, such as breast and prostate, also demonstrate that there is an inverse correlation between the presence of p53 mutations and bcl-2 expression [58,59].

3
Bcl-2 Family Members

Many bcl-2 family member proteins have now been identified (Fig. 1). These bcl-2 homologs can be broadly categorized as death antagonists and death agonists. The growing list of bcl-2 gene family members all share highly conserved domains referred to as bcl-2 homology domain 1 and 2 (BH1 and BH2) [60–62] or domains B and C, respectively [63,64]. These homology domains seem to be important for bcl-2 to form heterodimeric complexes with the family members and to carry out its anti-apoptotic function [61, 63, 65–67]. For example, mutations in BH1 and BH2 prevent bcl-2 from forming heterodimeric complexes with the bcl-2 homolog bax and can abrogate the survival function of bcl-2 [61]. The bcl-2 protein can also form homodimers with itself via its NH2 terminal region called the BH4 domain which spans residues 11 thru 33 [63].

3.1
Bax

Bax, bcl-2 associated X protein, is a death agonist member of the bcl-2 family of proteins. Discovered by co-immunoprecipitation with bcl-2, it was the first bcl-2 homolog to be identified [60]. The 4.5 Kb bax gene maps to 19q13.3–13.4 and is comprised of six exons [68]. It shares 21% identity and 43% similarity with bcl-2. The most conserved regions between the two molecules are within the BH1 and BH2 domains encoded by exons 4 and 5, respectively [60].

Multiple forms of bax protein can result from various splicing alternatives. The most prevalent form is bax-α, whose 1.0 Kb RNA encodes a 192 amino acid, 21 kD transmembrane protein. The 24 kD cytosolic bax-β lacks the transmembrane segment and is encoded by 1.5 Kb RNA transcript. A third form, bax-γ lacks the exon 2 and can undergo alternative splicing of intron 5 to yield 1.0 and 1.5 Kb RNA transcripts [69]. Yet another alternatively spliced form of bax, bax-δ, has the C-terminal transmembrane anchor as well as the BH1 and BH2 domains [68]. The functional role of these bax variants remains to be elucidated.

The bax gene promoter contains four p53 binding sites and the expression of bax is upregulated at the transcriptional level by p53 [70]. A temperature sensitive p53 mutant transfected into a myeloid cell line was associated with increased bax mRNA after shifting to the permissive temperature [71]. Also in cells obtained from p53 null mice, the level of bax proteins was found to be lower [72]. Moreover, following apoptosis induction by ionizing radiation, the bax mRNA was upregulated only in the cell line that possesses wild type p53 [71]. These data

Anti-Apoptotic Proteins

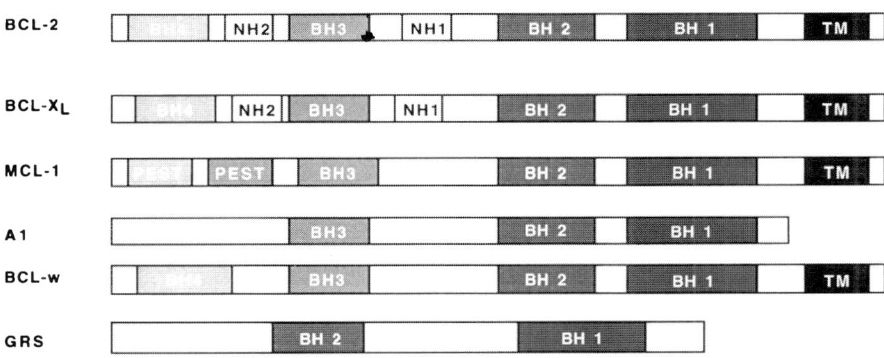

Pro-Apoptotic Proteins

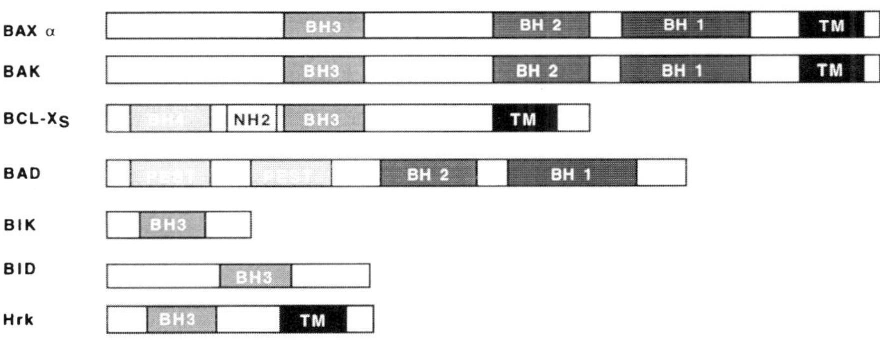

Fig. 1. Schematic depiction of the protein structures of the bcl-2 family members. BH1, BH2, BH3, and BH4 are the conserved homology domains. TM indicates the transmembrane domain, NH2 indicated the amino terminal domain, and the PEST domain represents the region which is correlated to an early response gene product and is associated with rapid protein turnover. GRS is grouped with the anti-apoptotic family members, however, its role in apoptosis is not currently known. Figure is not to scale

suggest that bax may function as a primary response gene in a p53 regulated apoptotic pathway [72]. However, thymocytes from the bax knockout mice were not diminished in their capacity to undergo apoptosis after γ-irradiation, a pathway driven by p53 [73]. Bax expression can also be modulated by other factors. The mRNA level has been shown to be downregulated in myeloid leukemia cell lines treated with IL-6 and/or dexamethasone [74]. The half life of bax mRNA can be increased in cell lines expressing higher levels of bcl-2 [75]. However, this increase in stability of bax mRNA by bcl-2 protein appears to be tissue specific.

Mutational analysis has shown that the BH1 and BH2 domains of bax are not required for heterodimerization with bcl-2, nor are the NH_2 terminal amino acids needed for bax homodimerization, unlike the homodimerization requirement for bcl-2. Rather, a stretch of amino acids spanning residues 59–101 in the BH3 domain was shown to be essential in both the homodimerization and heterodimer complex formation with bcl-2 [76]. Additionally, in contrast to bcl-2, bax can function in its monomeric form to accelerate cell death [77]. Bax can heterodimerize with other bcl-2-related proteins, including bcl-x_L, Mcl-1, and A1 [67]. The "rheostat" model has been proposed to explain the role of bcl-2 family member interactions in controlling cell death. This model suggests that the relative amounts of bcl-2 and bax may determine the susceptibility of a cell to undergo apoptosis [66]. According to this model, when bcl-2 is in excess, bcl-2/bax heterodimers predominate and cell death is inhibited. Conversely, when bax is in excess, bax homodimers predominate and the cell becomes susceptible to cell death induction following exposure to an apoptotic stimulus.

The tissue distribution of bax protein is more widespread than bcl-2 [78]. The immunohistochemical staining of murine tissues has revealed that the expression of bcl-2 and bax overlap in some tissues, and that bax is not always expressed at high levels in compartments marked by a high turnover rate. For example, bax, as well as bcl-2, are expressed in the thymic medulla but not in the thymic cortex, despite high numbers of cortical thymocytes which undergo apoptosis. Also, a high level of bax protein is observed in neurons, cells that have a long life. However, in certain tissues such as colonic epithelium, gastric glands, and secretory epithelial cells of prostate, bax expression corresponds to the cells that are susceptible to undergoing apoptotic cell death [78].

Evidence that apoptosis is not absolutely dependent on the expression of bax is also apparent from an analysis of the bax knockout mice. In these mice, the absence of bax is associated with either tissue specific hyperplasia or hypoplasia [73]. For example, there was an increase in number of resting mature B cells and thymocytes causing hyperplasia in the spleen and thymus. However, the male bax knockout mice were infertile due to atrophic testes, resulting from the abrogation of spermatogenesis [73].

Recent evidence suggests that bax may play a role as a tumor suppressor. Normally bax-α is expressed at high levels in breast tissue but is not detectable or is expressed at low levels in breast cancers [79]. Furthermore, in metastatic breast cancer, patients with reduced bax expression showed poor response to chemotherapy [80]. Transgenic mice have been generated which express a truncated form of the SV40 T antigen (Tg121) resulting in inactivation of the retinoblastoma protein but not p53. Tg121 mice bearing targeted disruptions of either the p53 gene or the bax gene exhibited an increased rate of brain tumor formation compared to Tg121 mice with intact p53 or bax genes [81]. Also, frequent frame shift mutations of bax were found in microsatellite mutator phenotype (MMP) colon adenocarcinomas, suggesting that the wild type bax gene may play a tumor suppressor role in colorectal carcinogenesis [82].

3.2
Bcl-x

Bcl-x was initially isolated from chicken lymphoid cells using a murine bcl-2 cDNA probe under low stringency conditions [83]. The bcl-x gene shares 44%

identity with bcl-2. Bcl-x was shown to interact with other members of the bcl-2 family in a manner similar to that shown for bcl-2 when analyzed by the yeast two-hybrid system [84]. Two human bcl-x cDNAs have been cloned [83]. Bcl-x$_L$ (long form) is a 31 kD protein, with an open reading frame of 233 amino acids. This form of bcl-x contains the BH1 and BH2 domains. The bcl-x$_L$ cDNA was found to be co-linear with the genomic sequence denoting the absence of mRNA splicing. Bcl-x$_S$ (short form) encodes a 170 amino acid, 19 kD protein. The carboxy-terminal 63 amino acids encoding the BH1 and BH2 domains are deleted from a 5′ splice site within exon 1 of the bcl-x gene [83]. A third alternative splice variant of bcl-x has been isolated from a murine cDNA library, bcl-x$_\beta$ [85]. Bcl-x$_\beta$ encodes a 209 amino acid protein that results from an unspliced first coding exon and lacks the carboxy-terminal 19 hydrophobic amino acids necessary for transmembrane insertion.

Both the level and pattern of expression of bcl-x differ from that of bcl-2. The levels of bcl-x expression are generally higher than bcl-2 in all tissues examined except for the lymph nodes where bcl-2 is predominant [78]. Bcl-x$_L$ is mainly expressed in the cells of the central nervous system, kidney, and bone marrow [85, 86]. Both bcl-x$_L$ and bcl-x$_S$, but not bcl-2, are expressed in CD34$^+$, CD38$^-$, lin$^-$ hematopoietic precursors [87]. However, the subcellular distribution of bcl-x protein is similar to bcl-2 in that it localizes to mitochondria and the nuclear envelope. This suggests that the function of the two proteins may be similar [85].

Further insight into the role of bcl-x during development was obtained from bcl-x deficient mice [88]. Heterozygous mice developed normally while homozygous, knockout mutants die at approximately day 13 of gestation. The bcl-x knockout embryos display extensive apoptosis involving post-mitotic neurons of the developing brain, spinal cord, dorsal root ganglia, and hematopoietic cells in the liver. Additionally, lymphocytes from bcl-x deficient mice showed diminished maturation. The life span of immature lymphocytes but not mature lymphocytes was shortened. This data indicates that bcl-x is required for the embryonic development of the nervous and hematopoietic systems.

Similar to bcl-2, bcl-x$_L$ was shown to confer resistance to apoptosis induction following growth factor deprivation. However, bcl-x$_S$ counteracted the ability of bcl-2 to block apoptosis [83]. Although bcl-x$_L$ and bcl-2 initially seemed to have the same functions, several observations suggest that biologically these two proteins are not completely overlapping. The tissue distribution of bcl-2 and bcl-x are not identical and the phenotypes of the corresponding knockout strains of mice are substantially different. Furthermore, it has been shown that WEHI-231 cells can be protected from apoptosis induced by surface IgM cross-linking by enforced bcl-x$_L$ expression while enforced bcl-2 expression exerts no such protective effect [89, 90].

The crystalline structure of bcl-x has expanded our insight into the potential mechanisms of function of bcl-2 family members [91]. Bcl-x structure was shown to consist of two central hydrophobic a helices surrounded by two amphipathic helices [91]. Interestingly, the conserved BH1, BH2, and BH3 domains were in spatial proximity and formed a hydrophobic cleft. This cleft is believed to form a binding site for other bcl-2 family members [91]. Evidence in favor of this hypothesis was provided when bcl-x and a 16 residue bak peptide derived from the BH3 domain were co-crystallized. The heterodimeric crystal structure revealed that the bak BH3 domain interacts with the hydrophobic cleft made by the BH1, BH2, and BH3 domains of bcl-x [92]. The crystal structure of bcl-x was also found to resemble the translocation domain of the diphtheria toxin and colicins [91]. This similarity in structure implies similarity in function and indicates that bcl-2 family members can be considered channel forming proteins capable of regulating the transmembrane trafficking of molecules involved in signaling cell death.

3.3
Bak

Bak (Bcl-2-homologous antagonist/killer) was first cloned from human heart and Epstein-Barr transformed human B-cell cDNA libraries [93–95]. There are three closely related bak genes (bak-1, 2, and 3) which are located on chromosome 6 (bak-1), chromosome 20 (bak-2) and chromosome 11 (bak-3). The bak genes contain at least three exons and span 6 Kb. Bak is a 211 amino-acid, 23 kD protein which shares 53% amino-acid identity with bcl-2. It possesses the same hydrophobic carboxy terminal domain as bcl-2 and bcl-x_L, which suggests that bak is an integral membrane protein. In contrast to bcl-2, bak is expressed at high levels in the kidney, pancreas, liver, and fetal heart, as well as adult brain [94]. Similar to bax in the intestine, bak expression is strongest in the cells in the luminal surface where most apoptosis is occurring. However, in a colorectal carcinoma cell line, only bak expression was shown to be modulated following apoptosis induction, indicating that bak may play a primary role in enterocyte apoptosis [96]. This contention is further supported by the observation that bak expression is reduced in colorectal adenocarcinoma samples. Therefore, a downregulation of bak may facilitate the accumulation of neoplastic cells in the early stages of colorectal tumorigenesis [97]. Bak was shown to accelerate cell death following IL-3 withdrawal [93, 94], but inhibits apoptosis induced by serum withdrawal and menadione treatment [93].

3.4
Bad

Bad (bcl-X_L/bcl-2 associated death promoter homolog) is a novel member of the bcl-2 family that was identified as a bcl-2 interacting protein using the yeast two-hybrid system [98]. The full-length bad cDNA sequence encodes a novel 204 amino acid protein with a predicted molecular weight of 22 kD. Bad shares only limited homology with known bcl-2 family members in the BH1 and BH2 domains. However, the functionally significant W/YGR triplet in BH1, the W at position 183, the WD/E at the exon junction in BH2, and the spacing between BH1 and BH2 domains is conserved. Unlike many other bcl-2 family members, bad does not contain a transmembrane anchor domain.

Bad was shown to heterodimerize with bcl-2 and bcl-x in vivo using co-immunoprecipitation. Bad's interaction with either bcl-2 or bcl-x can displace bax from the heterodimers. Significantly, this was shown to reverse the death repressor activity of bcl-x$_L$ but not of bcl-2. However, bad does not appear to interact with bax, Mcl-1, or A1, nor, apparently, does bad form homodimers [98]. Recent experiments have shown that bad may function in intracellular signal transduction pathways. Upon IL-3 stimulation of an IL-3 dependent hematopoietic cell line, bad becomes rapidly phosphorylated at two serine residues and is prevented from forming heterodimeric complexes with bcl-x$_L$. The phosphorylated bad is found to be complexed with 14-3-3, a phosphoserine binding protein which regulates protein kinases, and is sequestered in cytosol [99]. Therefore, only the non-phosphorylated bad is heterodimerized with the membrane bound bcl-x$_L$ and counters the anti-apoptotic activity of bcl-x$_L$. One of the models to explain the apoptotic activity of bad is that, in its non-phosphorylated form, bad binds to membrane associated bcl-x$_L$ which releases bax to enhance cell death [99]. Another link between the phosphorylation event and the apoptotic pathway was shown when it was found that, in vitro, bad is phosporylated by mitochondrial membrane targeted Raf-1, but not by the plasma membrane targeted Raf-1. Moreover, bcl-2 was shown to target Raf-1 to mitochondrial membrane which resulted in phosphorylation of bad and the subsequent enhancement of cell survival [100].

3.5
Mcl-1

Mcl-1 (human myeloid cell differentiation protein)was identified by differentially screening a cDNA library of the human myeloid leukemia cell line, ML-1, following induction by phorbol 12-myristate 13-acetate (TPA) [101]. Mcl-1 has also been detected in normal peripheral blood B cells after treatment with IL-4 and anti-IgM. Mcl-1 is an early response gene that reduces its expression immediately following differentiation induction [101, 102]. A study done using a yeast two-hybrid assay indicates that Mcl-1 interacts strongly and selectively with bax, but not with any other bcl-2 family members [67, 84].

Mcl-1 shares sequence homology with bcl-2 in the BH1 and BH2 domains and has a carboxy-terminal transmembrane anchor domain [102]. In addition, the Mcl-1 protein possesses PEST sequences [101], which correlate with the its role as an early response gene product [102]. The human Mcl-1 gene maps to chromosome 1 band q21 [103], an area often involved in chromosomal abnormalities in neoplastic and pre-neoplastic diseases [104–106].

Mcl-1 protects against apoptosis induced by constitutive expression of c-myc or bax [107]. However, in the 5AHSmyc cell line, Mcl-1 overexpression is not as effective as bcl-2 overexpression in preventing myc-mediated cell death [107]. It has been proposed that Mcl-1 may function as an alternative to bcl-2 in situations where bcl-2 cannot block apoptosis or in tissues lacking bcl-2 expression. For example, in normal peripheral blood B cells treated with agents which promote survival (IL-4, anti-μ, and TPA) or enhance rates of cell death (TGFβ1 and forskolin), upregulation of Mcl-1 correlates with cell survival and downregulation of Mcl-1 precedes cell death. In contrast, levels of bcl-2 expression are not modulated under the same experimental conditions [108].

Additionally, the tissue distribution of Mcl-1 and bcl-2 expression shows significant differences such as brain and spinal cord neurons in which bcl-2 predominates compared to skeletal muscle, cardiac muscle, cartilage and liver where Mcl-1 predominates over bcl-2 [109]. Similarly, Mcl-1 levels in normal lymph nodes are highest in germinal centers, where the rate of apoptosis is high. In contrast, bcl-2 is most intense in the mantle zone. It has been postulated that Mcl-1 temporarily blocks cell death until death suppressors such as bcl-2 are upregulated [110].

3.6
A1

A1 was identified by differentially screening a cDNA library of normal peripheral blood B cells after treatment with IL-4 and anti-IgM. The A1 cDNA was isolated from murine macrophages after GM-CSF induction of differentiation [111]. A1 is an early response gene that decreases its level of expression immediately following differentiation induction [111]. Yeast two-hybrid assays indicate that A1 interacts strongly and selectively with bax, but not with any other bcl-2 family member [67, 84]. A1 shares homology with bcl-2 in the BH1 and BH2 domains, but does not possess the carboxy terminal transmembrane domain [111].

The correlation of GM-CSF and LPS-induced differentiation with A1 upregulation suggests A1 could potentially function as a cell death suppressor [111]. Later reports has shown that A1 protects against TNF induced apoptosis in the presence of actinomycin D in a human microvascular endothelial cell line [112]. A1 could also inhibit ceramide induced cell death in these endothelial cells [112]. A1 expression displays a rather limited tissue distribution and appears to be confined to hematopoietic tissues, including helper T-cells, macrophages, and neutrophils [111].

3.7
Bfl-1

Bfl-1 (Bcl-2-related gene expressed in human fetal liver) was identified during a random cDNA sequencing project [113]. It was found to be homologous to bcl-2 family members with the highest homology to the A1 gene. The main region of homology was in the conserved BH1, BH2, and BH3 domains. Bfl-1 is mainly expressed in bone marrow while low levels of expression are detected in lung, spleen, esophagus, and liver. Bfl-1 mRNA was detected at relatively high levels in six out of eight stomach cancer tumors and metastasis when compared to normal stomach tissue from the same patients [113]. Bfl-1 protein suppresses apoptosis induced by p53 in the BRK cell line to the same extent as bcl-2 and bcl-x_L. Bfl-1 was also shown to cooperate with E1a in the transformation of primary rodent epithelial cells [114].

3.8
GRS

GRS was incidentally cloned during the cloning of fibroblast growth factor 4 (FGF-4) from a patient with chronic myelogenous leukemia [115]. The FGF-4 gene was truncated by a DNA rearrangement with a novel gene named GRS (Glasgow Rearranged Sequence) with a breakpoint 30 nucleotides downstream from the translation termination codon of FGF-4. The full length cDNA of GRS was then cloned from a human activated T cell cDNA library. The GRS cDNA is 824 nucleotides [116]. Sequence analysis of GRS revealed 71% identity to the murine A1 protein at the amino acid level.

Northern blot analysis showed a high level of expression of GRS in hematopoietic cells and to a lesser extent in lung and kidney [116]. GRS is also expressed in cell lines of hematopoietic origin including HL-60 (promyelocytic leukemia), Raji (Burkitt lymphoma) and K-562 (chronic myeloid leukemia). However GRS is not expressed in MOLT-4 T lymphoblastic leukemia and T cells prior to activation. The melanoma cell line G-361 also expressed high levels of GRS. GRS is localized to chromosome 15q24–25. This location positions GRS adjacent to t(15;17) region translocation frequently observed in acute promyelocytic leukemia. The GRS location also places it in the breakpoint described in Fanconi anemia that is associated with high incidence of acute leukemia.

3.9
Bid

Bid (BH3 interacting domain death agonist) was initially identified as a protein that interacts with both bcl-2 and bax proteins. The labeled bax and bcl-2 proteins were used to screen a λEX1ox expression library constructed from the murine T cell hybridoma line 2B4 [117]. Bid is a 23 kD, 195 amino acid protein. Sequence analysis of bid revealed that bid shares homology only with the BH3 domain of the bcl-2 family and that it lacks the carboxy terminus transmembrane hydrophobic domain. A human homolog of bid has also been identified. Human bid shares 72.3% sequence homology with the murine bid and has a 195 amino acid open reading frame [117].

In adult mouse tissue, bid is mainly expressed in the kidneys but is also present in brain, spleen, liver, testis, and lung [117]. Low levels of expression are detected in the heart and skeletal muscle. The mouse hematopoietic cell line, FL5.12, was also found to express high levels of bid. Subcellular fractionation has revealed that bid is predominantly localized to the cytosol (90%) with a small fraction in the membrane fraction [117].

Expression of bid in the IL-3 dependent FL5.12 cell line could induce a subtle but consistent enhancement of apoptosis following IL-3 withdrawal [117]. Bid inducible expression as well as transient transfections of bid in Rat-1 fibroblasts and Jurkat T-cells results in reducing cell viability to < 40% at 48 h [117]. Bid could also restore apoptosis in FL5.12 clones overexpressing bcl-2. The level of apoptosis was intermediate between the parental and bcl-2 overexpressing clones. The degree of cell death in all cases corresponded to the level of bid protein expression as detected by Western blot analysis. Bid induced apoptosis could be inhibited by zVAD-fmk, an irreversible inhibitor particularly effective against the CPP32-like subset of proteases. This suggests that bid induced cell death involves activation of CPP32-like proteases [117].

Bid interacts with both death agonists and antagonists members of the bcl-2 family. Bid can interact with bcl-2, bcl-x, and bax but does not form homodimers. Bid was unable to form trimolecular complexes with bcl-2 /bax heterodimers, suggesting that bid interacts with monomeric or homodimeric bcl-2 or bax. Several mutants of bcl-2, bax, and bid were examined to detect the regions of each molecule required for their interactions. The BH3 domain of bid was essential for interaction with bax and bcl-2. Differential specificity of these mutants was also detected as mutant (M 97 A, D 98 A) could bind bax but not bcl-2, mutant (G 97 A) could bind bcl-2 but not bax, while other mutants did not bind either protein. Noteworthy is that all BH3 mutants of bid were impaired in their ability to counter bcl-2 protection, including mutants that could still bind bcl-2. However, bid mutant (M97A, D 98 A) that can still bind bax but not bcl-2, retained its activity. Conversely, the BH1 domain of bcl-2 and bax were shown to be required for their interaction with bid. It is suggested that the α helix BH3 domain of bid interacts with the hydrophobic cleft contributed by the BH1 domain of bcl-x. This interaction might result in a conformational change in bid, bcl-2, or bax that signals cell death.

3.10
Bik

Bik (Bcl-2 interacting killer) is a novel bcl-2 family member that was detected when a human B-cell line cDNA library was used in a yeast two-hybrid screen for proteins that interact with bcl-2 [118]. Bik is a 160 amino acid protein and has a predicted molecular weight of 18 kD encoded by 928 bp cDNA and 1 Kb mRNA. Bik shares homology only within the BH3 domain of the bcl-2 family and has a carboxy terminal transmembrane hydrophobic domain. Bik was found to localize to the nuclear envelope and cytoplasmic membrane structures.

Transient co-transfection of bik and β-galactosidase expression plasmids in Rat-1 fibroblasts resulted in a dramatic reduction in the number of blue cells, consistent with reduced viability of bik transfected cells [118]. Co-transfection of bik and bcl-2, bcl-x, adenovirus E1B-19 kDa, or EBV-BHRF1 resulted in an increase in blue cell number, indicating the ability of these proteins to reverse cell death by bik. Deletion of the BH3 domain of bik resulted in loss of its proapoptotic activity. Bik induced apoptosis was shown to be inhibited by zVAD-fmk. However, CrmA could not inhibit bik induced cell death. This suggests that bik induced cell death involves selective activation of CPP32-like proteases [119].

Interactions between bik and other bcl-2 family members was examined using the yeast two hybrid system, GST-fusion protein capture on glutathione agarose beads, and transient co-transfection of tagged bik with other anti-apoptotic bcl-2 family members [118]. These in vitro and in vivo experiments revealed interactions between bik and bcl-2, bcl-x, adenovirus E1B-19 kDa, and EBV-BHRF1. Bik also interacts with bcl-x$_s$, a death promoting protein that lacks BH1 and BH2 domains. This suggests that Bik does not require BH1 and BH2 domain for its interaction with bcl-2 family members. Bcl-2 residues 43 – 48 and E1B-19 kDa residues 90 – 96 were shown to be essential for interaction with bik.

Noteworthy is that these residues are not within the conserved regions of bcl-2 family members.

3.11
Bcl-w

Bcl-w was cloned using degenerate primers to the conserved BH1 and BH2 domains in a low stringency PCR reaction [120]. These primers were used to amplify cDNA from mouse macrophage and mouse brain cell lines. The PCR product was then used to screen cDNA libraries from mouse brain, spleen, and myeloid cell lines. Bcl-w is a 22 Kb gene with a 3.7 Kb mRNA which encodes a 22 kD protein. Human bcl-w was then isolated from an adult human brain cDNA library. Bcl-w possesses the BH1, BH2, and BH3 domains. The human and mouse genes are 99% identical at the amino acid level and 94% at the nucleotide level.

Bcl-w mRNA is expressed at high levels in brain, colon, and salivary gland. Surprisingly, bcl-w expression is not detected in T and B lymphoid cell lines. However, mRNA was detected in myeloid cell lines of macrophage, mega-karyocyte, erythroid, and mast cell origin. Bcl-w also has a hydrophobic C-terminal transmembrane domain. The cytoplasmic localization of bcl-w is similar to that of bcl-2. Bcl-w resides in the central region of mouse chromosome 14 and human chromosome 14 at q11.2. Hematopoetic cell lines expressing bcl-w were resistant to apoptosis induction to the same extent as bcl-2 and bcl-x stable transfectants. However, bcl-w did not protect CH1 B lymphoma cells from CD95-induced apoptosis while bcl-2 and bcl-X_L were able to do so [120].

3.12
Harakiri

The Harakiri gene and its protein product Hrk was identified by a yeast two-hybrid screen of a HeLA cDNA library to detect proteins that bind to bcl-2 [121]. A nine-week human embryo cDNA library was used to obtain the full length Hrk cDNA. Hrk was detected as a 716 bp cDNA that was confirmed by the Northern blot analysis using both human and mouse tissue as 0.7 Kb mRNA. The cDNA encodes an open reading frame of 91 amino acids. Hrk shares homology with bcl-2 family member BH3 domain; however, the rest of the protein has no significant homology to any other protein or bcl-2 family. A region of 28 hydrophobic amino acids that may serve as a membrane-spanning domain was also identified at the COOH-terminus of Hrk.

Northern blot analysis demonstrates high levels of Hrk expression in all lymphoid tissues examined including the bone marrow and spleen. Hrk is also expressed in the pancreas and at low levels in the kidney, liver, lung, and brain [121]. Hrk was seen as a cytosolic granular staining by confocal microscopy of transiently transfected cells with flagged Hrk. This staining is similar to the previously reported localization of bcl-2 and bcl-x.

Transient transfections of Harakiri in 293 T cells, HeLa, and FL5.12 progenitor B-cells resulted in a dramatic decrease in cell viability by 36 h post-transfection. However co-expression of bcl-2 and bcl-x could inhibit the death promoting activity of Hrk. Interestingly, Hrk appears to interact only with bcl-2 and bcl-x$_L$ but not with the other pro-apoptotic family members bax, bak, and bcl-x$_S$. Deletion mutants of Hrk lacking 16 amino acids including the BH3 domain were unable to interact with bcl-2 and bcl-x. This mutant also failed to induce cell death in 293 T cells. Deletion analysis has also revealed the requirement of BH1 and BH2 domains of bcl-2 and bcl-x to interact with Hrk.

4
Interactions of Bcl-2 Family Members and Mechanisms of Function

4.1
Interactions Keep a Family Close

There is now voluminous scientific literature related to cell death research. The research findings reported in this literature are frequently at variance and in apparent conflict. What we must strive toward now is an integrated representation of how various cell death regulatory molecules interplay to effect the apoptotic process. It is, perhaps, misleading to consider that such an accomplishment would not necessarily require a tour de force, particularly since the morphological manifestations of apoptosis are relatively invariant.

The roles of the bcl-2 family of cell death effector proteins, which this review considers, are no less certain. Below, we attempt to place the bcl-2 family in some perspective with respect to their molecular features. Owing to the fast pace of research in this field, perhaps subsequent reviews will be able to arrive at a more uniform account of their biochemical role in apoptosis.

One of the reasons for our modest understanding of the mechanisms by which bcl-2 homologs execute their cellular roles stems from a lack of identifiable sequence motifs in the bcl-2 family which would implicate a mechanism of action. What have been defined, however, are shared domains designated as bcl-2 homology domain 1, 2, 3, and 4. The BH1 domain spans amino acid residues 136–155 of the bcl-2 protein, BH2 spans resides 187–202, BH3 spans resides 93–107, and BH4 spans residues 10–30. The BH3 domain, for its part, appears to be involved in selective interactions between bcl-2 family members. The BH3 domain appears to be required for the death promoting activity of bax and bak which are also required for their interaction with the two death-repressing members, bcl-2 and bcl-X$_L$ [76, 93]. The BH1 and BH2 domains serve equally critical functions. The creation of point mutations in either domain can effectively abolish the death repression function of bcl-2 [61]. Recent evidence suggests, however, that the formation of heterodimers is not required for function of family members [122]. These same BH1 and BH2 domain mutants of bcl-2 fail to heterodimerize with bax, although they do homodimerize well [61]. Some of the most compelling evidence that the BH3 motif represents a "death domain" comes from studies of bid [117]. Bid possesses only the BH3 domain, lacks the carboxy-terminal signal-anchor segment, and localizes to both cytosolic and membrane compartments. Importantly, ectopic expression of bid abrogates the pro-survival effect of bcl-2. Additionally, expression of bid, without another death stimulus, induces

ICE-like proteases and apoptosis. An intact BH3 domain of bid was required to bind the BH1 domain of either bcl-2 or bax.

The BH4 domain, which is located at the amino-terminus, has been far less characterized. To date, it has been reported that deletion of the BH4 domain of bcl-2 nullifies anti-apoptotic function and homodimerization, but does not impair bcl-2/bax heterodimerization [123]. There is some evidence which indicates that the BH4 domain may mediate interactions of bcl-2 family member protein with non-bcl-2-related proteins such as calcineurin [124]. Thus the BH4 domain may serve as a tethering domain that bridges bcl-2 and bcl-2-related proteins to important signal transduction proteins.

4.2
Models Involving Familial Interactions

Perhaps, at its simplest level, the expression of various bcl-2-related proteins may determine whether a cell responds to an applied stress by initiating a cell death program or surviving. However, another hypothesis, that has substantial experimental evidence based on a mutational analysis of the BH domains, suggests that the cellular response to injury may be a function of the multiple heterodimerization and homodimerization states between members of this protein family. This model, commonly known as the "rheostat model", has been advocated by Dr. Stanley Korsmeyer's group [60, 66]. In this scenario, the relative levels of dimerization partners available shifts the balance of cell fate in favor of viability (e.g., bcl-2/bcl-2 homodimers favoring cell survival) or cell death (e.g., bax/bax homodimers favoring cell death) following exposure to an appropriate stress. This ability of bcl-2-related proteins to hetero- and homodimerize in vivo is perhaps one of the most important features of the family. Complicating the picture further are reports of the ability of several bcl-2 family members to interact physically with several signaling protein complexes containing p21 ras [125], Raf-1 kinase [126], and p23 R-ras proteins [127]. Another feature is the conservation of a hydrophobic membrane targeting sequence in the carboxy terminal tail of most members of the bcl-2 family. The targeting domain most likely ensures that the various members are correctly routed to the appropriate intracellular organelle. Perhaps this routing domain ensures that the various bcl-2-related proteins are localized in close proximity to secure proper physical interactions should the appropriate stress be "detected".

4.3
Stress Signaling Pathways Converge on Bcl-2 Family Members

One certain feature of apoptosis signal transduction is that a cell must quickly integrate the balance of multiple pro-apoptotic and anti-apoptotic signals, then respond appropriately. How the different cells integrate this stress and how they "choose" to respond to divergent stresses can be modulated by expression of bcl-2 gene family members. In this regard, bcl-2 and certain other bcl-2 homologs may actually function as molecular "sensors" of the extent of cellular damage. The mechanism by which bcl-2 and its homologs "integrate" and/or relay stress signals is at best vaguely understood, despite the rapid progress that is being made in this area. It is reasonable to assume that multiple, ostensibly divergent stress signals may converge on bcl-2 and other bcl-2 family member homologs.

The goal of this review is to define some points of stress signaling convergence that might help explain the function of bcl-2. The ceramide/SAPK/JNK and NFκB pathways, in particular, will be emphasized as pathways intersecting with bcl-2. We will also speculate on some recent findings that indicate bcl-2 family proteins may generally regulate the transmembrane trafficking of molecules implicated in the mediation of apoptosis.

4.4
Common Molecular Themes of Divergent Stress Signals

Bcl-2 has the unique ability to inhibit multiple cell death signaling pathways such as those mediated by TNF-α [128–131], activation of cysteine proteases such as ICE [132], reactive oxygen species (ROS) [66, 133, 134], γ- and UV-radiation [55, 135], heat shock [136], p53 [55], c-myc [51], calcium [137–139], chemotherapeutic agents [140, 141], androgen withdrawal [142], neurotrophin withdrawal [143], and anti-CD3 receptor clustering [144].

These observations prompt several questions: (a) does a common molecular event exist for these cell death pathways?; (b) do pro-survival bcl-2 family members execute their cellular role(s) at this level or are divergent mechanisms used by bcl-2 to attenuate different stresses?; and (c) do the pro-survival members of the bcl-2 family serve as genuine stress signaling effectors that can nullify or attenuate a given stress signal(s) in the case of anti-apoptotic members, or amplify the signal(s) in the case of pro-apoptotic members?

Many of the intimated diverse stress pathways potently and rapidly activate the "Sphingomyelin Cycle" and subsequent events which ensue in the nucleus [145]. The activation of the Sphingomyelin Cycle, which is mediated by a membrane sphingomyelinase (Smase) and catalyzes the cleavage of plasma membrane sphingomyelin resulting in ceramide generation, has been documented for TNF-α, UV- and γ-radiation, heat shock, and peroxides induced apoptosis [146]. Ceramide, in turn, can strongly activate the JNK/SAPK signaling pathway [146]. The involvement of the JNK/SAPK pathway in mediating apoptosis signaling is underscored by the ability of dominant negative loss of function mutants in c-jun (a principal substrate of JNK/SAPK) or an upstream activator of JNK/SAPK (JNK kinase) to block ceramide-induced apoptosis [146]. However, the exact role of the JNK/SAPK pathway in signaling apoptosis is controversial since others have noted a lack of JNK/SAPK involvement in the TNFR1-activated apoptosis signaling pathway [147].

It is interesting to note that the NFκB survival signaling pathway can be activated, in some cells, by cell membrane permeable analogs of ceramide. This observation may provide a link that integrates the sphingomyelin cycle for cell stress with the NFκB-cell survival signaling pathway [148]. Therefore, it is reasonable to consider that the ceramide-activated stress pathway may represent a common denominator which may provide insight into the basis of bcl-2 function given the fact that bcl-2 can suppress apoptosis induced by these stresses.

4.5
The NFκB Distress Response: Role of Bcl-2

It has been proposed that NFκB may play a role in the regulation of apoptosis because it is activated as a result of multiple stress responses [149]. Additionally, bcl-2 has the ability to suppress apoptosis which is associated with NFκB-mediated stress responses. Therefore, it would be of interest to examine the relationship between bcl-2 and NFκB. The NFκB signaling pathway is initiated by stress signal-induced phosphorylation of IκBα which may precede in many instances via a MEKK-dependent pathway [150]. This results in the simultaneous dissociation of IκBα from the IκBα/NFκB complex, the ubiquitinization, and subsequent degradation of IκBα. The end result is rapid nuclear import of activated NFκB heterodimers (e.g., Rel A: p65 – p50) [151]. Nuclear NFκB cooperates with other transcription factors such as Sp1, ATF-2, c-jun, and ETS. Consensus κB binding sites have been discovered in the promoters of many common genes implicated in the mediation of apoptotic cell death including c-myc, ICE, and p53 [151]. This suggests that the NFκB signal, once decayed due to resetting of the cytoplasmic levels of IκBα, may be propagated by waves of downstream gene activation. Recently, bcl-2 has been shown to modulate NFκB activation following cell death induction mediated by Sindbis virus (SV) either directly or indirectly [149]. In this system, it was demonstrated that apoptosis is strictly dependent on an NFκB signaling mechanism since NFκB DNA binding site "decoys" (synthetic NFκB binding site double stranded oligonucleotides) inhibited SV-induced apoptosis. Interestingly, bcl-2 appeared to block the nuclear import of NFκB following apoptosis induction by SV. These data argue for a pro-apoptotic role of the NFκB signaling pathway. Compelling evidence has recently been presented, however, demonstrating that genetic disruption of the NFκB signaling pathway using dominant-negative acting mutants of IκBα that can sensitize cells to TNF-α-induced apoptosis [152, 153]. These reports assert that the NFκB signaling pathway represents an adaptive pro-survival response to acute cell stress. Interestingly, ectopic expression of bcl-2 in human prostatic carcinoma cells protects against the loss of NFκB signaling in response to TNF-α, indicating that bcl-2 functions downstream of the NFκB signaling pathway [152]. Bcl-2 had no effect on the kinetics of IκBα degradation or p65 RelA nuclear import following TNF-α stimulation [152]. Consistent with these observations is the recent demonstration that bcl-X_L can block apoptosis by PDTC, an inhibitor of IκBα degradation, in WEHI B cell lymphoma cells that constitutively express NFκB [154]. The mechanism by which bcl-2 and bcl-X_L can signal cell survival in the absence of NFκB remains uncertain.

Irrespective of the basic mechanism, the potential for bcl-2 and bcl-X_L to supplant for the loss of an NFκB-generated survival signal(s) highlights a potential problem to NFκB blocking strategies recently proposed to augment TNF-α mediated cytotoxicity in tumor cells.

4.6
Function by Location

Bcl-2, by virtue of its carboxy-terminal hydrophobic anchor sequence, localizes to three membranous compartments: the mitochondrial membrane, ER, and nuclear envelope [155, 156]. Interestingly, it is these intracellular organelles where reactive oxygen species (ROS) are also generated. It has been demonstrated that bcl-2 can block apoptosis induced by γ-radiation and TNF-α, two pathways known to involve the generation of ROS; however, the regulation of ROS and lipid peroxidation by bcl-2 is still unclear [55, 128–131].

Cell death signaling events are commonly mediated by alterations in intracellular Ca^{2+} [137–139]. A common feature of oxidative stress is the initiation of mitochondrial calcium cycling [157]. Within the mitochondria, hydrogen peroxide causes calcium release which in turn results in lipid peroxidation [158]. If calcium release is inhibited, lipid hydrolysis is reduced [159]. The subcellular localization of bcl-2 also corresponds to sites of intracellular Ca^{2+} storage. It has recently been demonstrated that bcl-2 participates in the regulation of cytoplasmic [138, 139] and intranuclear calcium [139] following apoptosis induction. Additionally, TNF-α induced apoptosis was also blocked by inhibition of calcium uptake in the mitochondria [160] and bcl-2 blocks TNF-α induced apoptosis.

Generation of ceramide to ceramide 1-phosphate by ceramide kinase has also been shown to be calcium dependent [161]. Three sphingomyelinases (SMase) exist which also hydrolyze sphingomyelin (SM) to ceramide. The acidic SMase functions optimally at the acidic pH found within the lysosomes. The second SMase also functions at acidic pH but also requires Zn^{2+} as a cofactor. A neutral SMase, which is membrane bound, functions at neutral pH, and can be activated by TNF-α [162]. It is conceivable that bcl-2 may reduce ceramide production by inhibiting SM hydrolysis by inhibiting intracellular increases in Ca^{2+}.

4.7
Bcl-2 as a Sentry of Nuclear Protein Import

Using immunogold transmission electron microscopic studies, bcl-2 appears to be distributed in a patchy distribution resembling nuclear pore complexes (NPCs) [155, 156]. This observation raises the possibility that NPC-associated bcl-2 regulates the biophysical properties of NPCs in such a manner as to regulate nuclear trafficking events. In consideration, it is notable that c-myc and bcl-2, which compliment to block p53-dependent apoptosis, have been observed to cooperate to block p53 nuclear import in G_1 of the cell cycle [163]. Moreover, another report indicates that the nuclear translocation of two cyclin A-dependent kinases, Cdc2 and Cdk2, is inhibited by enforced bcl-2 expression [164]. It is uncertain how bcl-2 imposes the block to nuclear import. Two possibilities need to be addressed: (a) is the block to import achieved by a bcl-2-regulated process at the level of the nuclear pore or (b) is the bcl-2-regulated block to nuclear import achieved more upstream via an interruption of cytoplasmic events necessary for import? Such signals might include bcl-2-regulated alterations in the binding of chaperone proteins, phosphorylation, or other posttranslational modifications.

The rationale for exploring a nuclear "gatekeeper" role for bcl-2 is rather direct insofar as many cellular alterations, both pro-survival and pro-apoptotic, occurring during apoptosis take place in the nucleus. For instance, apoptosis induction caused by ROS, UV- and γ-radiation, adriamycin, and daunorubicin result in nuclear alterations including oxidation, cleavage, intercalation, and thymidine-dimer formation of DNA. Additionally, caspases become activated and degrade nuclear lamins, small nuclear ribonuclear proteins (snRNPs), and DNA repair enzymes such as poly ADP-dependent polymerase (PARP) and DNA-dependent protein kinase (DNA-PK) [165, 166]. Likewise, certain pathways for apoptosis, such as apoptosis induced by ionizing radiation, depend on the nuclear function of wild type p53, a tumor suppressor protein with potent growth arresting activity [55].

To highlight one intriguing possibility, it is feasible that bcl-2 moderates the nuclear import of p53 by binding to a mutual dimerization partner for p53 and bcl-2, designated Bbp/53 bp [167]. Bbp/53 bp, was cloned in a yeast two-hybrid screen for bcl-2 binding proteins using a carboxy-terminal truncated version of bcl-2. A purified Bbp/53 bp fusion protein appears to interact with bcl-2 and p53 in a mutually exclusive manner [167]. The function of Bbp/53 bp is unknown at present, leaving the possibility that cytosolically localized Bbp/53 bp serves a co-import chaperone-like function for p53.

Another possibility involves the ability of bcl-2 to bind proteins involved in protein modification. To illustrate, it has recently been demonstrated that bcl-2 can modulate the nuclear import of NF-AT by sequestering calcineurin [124]. By binding calcineurin in a bax-inhibitable manner, via its BH4 domain, bcl-2 sequesters calcineurin in a phosphatase-active complex that is unable to interact with NF-AT, a prerequisite for NF-AT nuclear import following stimulation of cells with a calcium ionophore [124]. Importantly, the BH4 domain of bcl-2 is located at its amino-terminus and is also required for its interaction with Raf-1 kinase. The interaction of Raf-1 via the BH4 domain of bcl-2 is responsible for targeting Raf-1 to the mitochondrial membrane [100]. The potential for competitive interaction of a protein kinase, Raf-1, and a protein phosphatase, calcineurin, with the BH4 domain of bcl-2 raises the possibility for molecular interplay between two signal transduction pathways at the level of bcl-2.

4.8
Bcl-2 Family Members as Ion Channels

One potential mechanism of bcl-2 bioactivity is the regulation of mitochondrial function(s) during apoptosis. Interestingly, bcl-2 preferentially localizes to sites within the mitochondrial membrane where the inner and outer leaflets are closely approximated [156]. These sites are associated with mitochondrial transmembrane trafficking. It has recently been shown that bcl-2 can impose a block to the induction of the mitochondrial permeability transition (PT) that accompanies a reduction in the mitochondrial transmembrane potential and correlates with a bcl-2 mediated impairment of PT-associated apoptosis [168]. How bcl-2 regulates the PT during apoptotic signals is unclear. Perhaps bcl-2 interacts directly with the PT pore. Alternatively, bcl-2 has been reported to enhance oxidative phosphorylation [169] which might reduce the probability of a PT event.

Another example of studies indicating that bcl-2 can regulate membrane events has been documented for prostate carcinoma cells stressed by thapsigargin (TG), a common chemotherapeutic agent and inhibitor of the ER-associated Ca^{2+} pump. In this study, ectopic expression of bcl-2 effectively squelched the depletion of Ca^{2+} stores from the ER following cell death induction by TG [139]. Bcl-2 transfected cells imaged by confocal microscopy for intracellular Ca^{2+} distribution using a Ca^{2+}-binding dye Fluo-3 appeared to establish a Ca^{2+}-gradient across the nuclear envelope following cell death induction by TG. Control cells responded to TG by releasing Ca^{2+} from their ER pools and redistribution of the released Ca^{2+} throughout the cell including the nucleus [139].

Together these observations suggest that one activity of bcl-2 may be to regulate selectively the transmembrane trafficking of cell death and possibly survival effector molecules. In this regard, some of the most compelling evidence that a bcl-2-related protein can regulate transmembrane trafficking comes from some recent studies of bcl-X_L. In this study it was demonstrated that purified bcl-X_L can form cation-selective ion channels in synthetic lipid bilayer membranes comprised of negatively charged lipids at physiological pH [170]. This data has broad implications with respects to the molecular basis of bcl-2 cell death regulation. The localization of bcl-2 family membrane proteins to intracellular compartments might then be anticipated to effect profoundly various cellular responses by selectively regulating membrane permeability. One possibility that must be explored in detail, is how the interaction of bcl-X_L with other bcl-2 family member proteins might control its ion channel forming potential.

Perhaps the membrane targeting of bcl-X_L reflects its capacity to regulate other processes related to mitochondria. One very interesting possibility regards the release of cytochrome C during some types of pharmacologically-induced apoptosis. In support of this notion are recent reports that bcl-2 can modulate the release of cytochrome C in response to divergent stress stimuli including staurosporin, etoposide, and UVB [171, 172]. Addition of purified cytochrome C bypassed the inhibitory effect of bcl-2 and induced a DEVD-specific caspase activity and nuclear apoptotic changes [171, 172]. This scenario may reflect a generalized cell death paradigm, in which a pro-survival molecule functions to block the release of a pro-death molecule from sequestration within a membranous organelle.

4.9
Signaling Complexes Containing Bcl-2

One exciting possibility to define the function of bcl-2 lies in reports that bcl-2 and bcl-2/bag-1 can be recruited into multimeric complexes with Raf-1 kinase [173] and two GTPases, p21 ras [125] and R-ras [126]. The interaction of Raf-1 via the amino-terminal BH4 domain of bcl-2 is responsible for targeting Raf-1 to the mitochondrial membrane [100]. In this system, active Raf-1 fused with targeting sequences from an outer mitochondrial membrane protein not only protected cells from apoptosis but also phosphorylated bad, a potent death promoting relative of bcl-2. Moreover, a kinase-inactive Raf-1 mutant abrogated apoptosis suppression by bcl-2, suggesting that the pro-survival signaling function of bcl-2 requires active kinases. The involvement of the bcl-2/Raf-1 kinase signaling pathway has important ramifications given that the ras/Raf/MAPK may directly oppose the p38 MAPK stress signaling pathway [173].

It has also recently been reported that bcl-2 can be detected in anti-Ras immunoprecipitates in Jurkat cells stimulated with phorbol esters under conditions that would downregulate PKC activity [125]. Using metabolic labeling with radiolabeled phosphate, the bcl-2 in the bcl-2/ras complex was found to be substantially more phosphorylated. Co-stimulation with staurosporin, a serine/threonine kinase inhibitor, abolished bcl-2 phosphorylation in this system and bcl-2-mediated anti-apoptosis signaling [125], suggesting a functional link between bcl-2 phosphorylation and the suppression of apoptosis. Given that bcl-2 can interact with bag-1, bag-1 and bcl-2 can interact with Raf-1 kinase, bcl-2 can bind to p21 ras, and p21 ras is an upstream activator of raf activity, it could be speculated that bcl-2 may be recruited into multimeric signaling complexes in certain cell types under specific stress conditions.

Additionally, using site-specific mutants of bcl-2 modified at potential PKC phosphorylation sites (ala for ser substitutions) indicates that phosphorylation of serine 70 of bcl-2 is necessary to ensure proper bcl-2 antagonism of cell death following IL-3 withdrawal of an IL-3-dependent cell line [174]. Phosphorylation of serine 70 was also essential to block etoposide-induced apoptosis. Surprisingly, the loss-of-function S70 A mutant of bcl-2 could still dimerize with bax in vivo, suggesting that bcl-2 phosphorylation activates the anti-apoptotic function of bcl-2 and not the dimerization interaction with bax per se [174].

4.10
Bcl-2 Regulation of Intracellular Protease Pathways

Another prevalent molecular theme has emerged that characterizes cells undergoing apoptosis, namely the strong and sustained activation of multiple intracellular proteases. The activation of intracellular proteases perhaps reflects a commitment phase in the execution of the apoptosis program whereby cells systematically destroy important cellular structural proteins and enzymes.

Much of our understanding of the role of proteases in cell death regulation is owed to Dr. Robert Horvitz and his colleagues. The work of this group has centered on developmentally regulated cell death in the nematode *C. elegans*. The work of this group initially identified three genes whose expression were intimately associated with the execution of certain cells during *C. elegans* development including ced-3 [10], a gene that encodes a cysteine protease homologous to ICE. Embryos with the ced-3 mutation fail to undergo deletion of certain critical cells during embryogenesis [10].

The caspase family of proteases now includes 11 members [165, 166]. One characteristic of this proteases family is that all members cleave protein substrates containing the YVAD (for ICE-like) or DEVD (for CPP32b-like) tetra-peptides after the aspartate (D) at the P1 position. All are synthesized as pro-enzymes requiring cleavage to be catalytically activated.

The involvement of these proteases in signaling apoptosis is taken from multiple sources which have been recently reviewed [12, 165, 166]. First, ectopic expression of ICE-related proteases is a sufficient signal to induce apoptosis in a number of cell types. In addition, ICE-like proteases can induce nuclear changes characteristic of cells undergoing apoptosis in in vitro assays. Finally, specific tetrapeptide inhibitors (DEVD and YVAD) or other inhibitors of ICE-related proteases (e.g., baculovirus p35 protein) can block apoptosis signaling by a variety of cellular stresses.

Since many of the cell stresses that activate ICE- or CPP32b-related proteases are also blocked by ectopic expression of bcl-2 and other bcl-2-related proteins, it would follow that pro-survival signaling bcl-2 family proteins might function downstream of these proteases. Likewise, it might follow that pro-apoptotic bcl-2-related proteins such as bax, bcl-X$_s$, and bad might function upstream of these proteases to signal their activation.

It has been shown that ced-9 and its mammalian homolog bcl-2 could both interact with and inhibit the activation of ced-4, a pro-apoptotic protein that activates the ced-3 protease or its mammalian counterparts, the ICE and FLICE proteases [13]. These data indicate that ced-4 plays a central role in the cell death pathway, biochemically linking ced-9 and the bcl-2 family to ced-3 and the ICE family of pro-apoptotic cysteine proteases.

5
Conclusions

The mechanisms of programmed cell death are far from being completely eluci-dated. At present, many different factors such as protease activation, DNA cleav-age, and calcium signaling are known to participate in apoptosis. The placement of bcl-2 and bcl-2 family members in cell death regulatory pathways is now being elucidated. It is now known that bcl-X$_L$ can form ion channels and it may be that other bcl-2 family members function in a similar manner. The specific interactions that bcl-2 family proteins have with various signaling molecules and within the bcl-2 family itself are active areas of investigation.

Acknowledgements. This work was supported in part by CaP CURE, The Association for the Cure of Cancer of the Prostate, NIH grant CA62,597 and CA68,233, and American Cancer Society grant DHP-156.6.

6
References

1. Saunders JW (1966) Science 154:604
2. Wyllie AH, Kerr JFR, Currie AR (1980) Int Rev Cytol 68:251
3. Kerr JFR, Wyllie AH, Currie AR (1972) Br J Cancer 26:239
4. Bakhshi A, Jensen JP, Goldman P, Wright JJ, McBride OW, Epstein AL, Korsmeyer SJ (1985) Cell 41:899

5. Cleary ML, Sklar JCM (1985) Proc Natl Acad Sci USA 82:7439
6. Cleary ML, Smith SD, Sklar J Cell (1986) 47:19
7. Tsujimoto Y, Cossman J, Jaffe E, Croce CM (1985) Science 228:1440
8. Tsujimoto Y, Croce CM (1986) Proc Natl Acad Sci USA 83:5214
9. Ellis HM, Horvitz HR (1986) Cell 44:817
10. Yuan J, Shaham S, Ledoux S, Ellis HM, Horvitz HR (1993) Cell 75:641
11. Patel T, Gores GJ, Kaufmann SH (1996) FASEB J 10:587
12. Fraser A, Evan G (1996) Cell 85:781
13. Chinnaiyan AM, O'Rourke K, Lane BR, Dixit VM (1997) Science 275:1122
14. Nguyen M, Milar DG, Yong VW, Korsmeyer SJ, Shore GC (1993) J Biol Chem 268:25265
15. Hengartner MO, Ellis RE, Horvitz HR (1992) Nature 356:494
16. Hengartner MO, Horvitz HR (1994) Cell 76:665
17. Vaux DL, Weissman IL, Kim SK (1992) Science 258:1955
18. Novack DV, Korsmeyer SJ (1994) Am J Path 145:61
19. LeBrun DP, Warnke RA, Cleary ML (1993) Am J Path 142:743
20. Chandler D, El-Naggar AK, Brisbay S, Redline RW, McDonnell TJ (1994) Human Pathol 25:789
21. Yachnis AT, Powell SZ, Olmsted JJ, Eskin TA (1997) J Neuropath Exp Neurol 56:186
22. Lu QL, Poulsom R, Wong L, Hanby AM (1993) J Pathol 169:431
23. Veis DJ, Sorenson CM, Shutter JR, Korsmeyer SJ (1993) Cell 75:229
24. Kamada S, Shimono A, Shinto Y, Tsujimura T, Takahashi T, Noda T, Kitamura Y, Kondoh H, Tsujimoto Y (1995) Cancer Res 55:354
25. Nakayama K, Negishi I, Kuida K, Sawa H, Loh DY (1994) Proc Natl Acad Sci USA 91:3700
26. Hockenbery DM, Zutter M, Hickey W, Nahm M, Korsmeyer SJ (1991) Proc Natl Acad Sci USA 88:6961
27. Nunez G, Hockenbery DM, McDonnell TJ, Sorensen CM, Korsmeyer SJ (1991) Nature 353:71
28. McDonnell TJ, Troncoso P, Brisbay S, Logothetis C, Chung LWK, Hseih J, Tu S, Campbell ML (1992) Cancer Res 52:6940
29. Bishop JM (1991) Cell 64:235
30. Nowell PC, Hungerford DA (1960) Science 132:1497
31. Rowley JD (1973) Nature 243:290
32. Manalov G, Manolova Y (1972) Nature 237:33
33. Zech L, Haglund M, Nilsson K, Klein G (1976) Int J Cancer 17:47
34. Shtivelman E, Lifshitz B, Gale RP, Canaani E (1985) Nature 315:550
35. Dalla-Favera R, Bregni M, Erickson J, Patterson D, Gallo RC, Croce CM (1982) Proc Natl Acad Sci USA 79:7824
36. Taub R, Moulding C, Battey J, Murphy W, Vasicek T, Lenoir CM, Leder P (1984) Cell 36:339
37. Nishikura K, ar-Rushdi A, Erikson J, Watt R, Rovera G, Croce CM (1983) Proc Natl Acad Sci USA 80:4822
38. Langdon WY, Harris AW, Croy S, Adams JM (1986) Cell 47:11
39. Tsujimoto Y, Finger LR, Yunis J, Nowell PC, Croce CM (1984) Science 226:1097
40. Vaux DL, Cory J, Adams JM (1988) Nature 335:440
41. Hockenbery DM, Nunez G, Milliman C, Schreiber RD, Korsmeyer SJ (1990) Nature 348:334
42. McDonnell TJ, Deane N, Platt FM, Nunez G, Jaeger U, McKearn JP, Korsmeyer SJ (1989) Cell 57:79
43. McDonnell TJ (1993) Mol Carcinogen 8:209
44. Seto M, Jaeger U, Hockett RD, Graninger W, Bennett S, Goldman P, Korsmeyer SJ (1988) EMBO J 7:123
45. Zutter, M, Hockenbery, D, Silverman, GA, Korsmeyer SJ (1991) Blood 78:1062
46. deJong D, Prins F, van Krieken HH, Mason DY, van Ommen GB, Klein PM (1992) Curr Top Microbiol Immunol 182:287
47. Merino R, King L, Veis D, Korsmeyer SJ, Nunez G (1994) EMBO J 13:683
48. Gratiot-Deans J, King L, Jurka A, Nunez G (1993) J Immunol 151:83

49. McDonnell TJ, Nunez G, Platt FM, Hockenbery D, London L, McKearn JP, Korsmeyer SJ (1990) Mol Cell Biol 10:1901
50. McDonnell TJ, Korsmeyer SJ (1991) Nature 349:254
51. Marin MC, Hsu B, Stephens C, Brisbay S, McDonnell TJ (1995) Exp Cell Res 217:240
52. McDonnell TJ (1993) Transgene 1:47
53. Gawerky CE, Haluska FG, Tsujimoto Y, Nowell PC, Croce CM (1988) Proc Natl Acad Sci USA 85:8548
54. Bissonnette RP, Echeverri F, Mahboubi A, Green DR (1992) Nature 359:552
55. Marin MC, Hsu B, Meyn RE, Donehower LA, El-Naggar AK, McDonnell TJ (1994) Oncogene 9:3107
56. Wang Y, Szekely L, Okan I, Glein G, Wiman KO (1993) Oncogene 8:3427
57. Chiou S-K, Rao L, White E (1994) Mol Cell Biol 14:2556
58. Silvstrini R, Veneroni S, Daidone MG, Benini E, Boracchi P, Mezzetti M, DiFronzo G, Rilke F, Veronesi U (1994) J Natl Cancer Inst 86:499
59. McDonnell TJ, Navone NM, Troncoso P, Pisters LL, Conti C, von Eschenbach AC, Brisbay S, Logothetis CJ (1997) J Urology 157:569
60. Oltvai ZN, Milliman CL, Korsmeyer SJ (1993) Cell 74:609
61. Yin XM, Oltvai ZN, Korsmeyer SJ (1994) Nature 369:272
62. Yin XM, Oltvai ZN, Korsmeyer SJ (1995) Curr Top Microbiol Immunol 194:331
63. Hanada M, Aime-Sempe C, Sato T, Reed JC (1995) J Biol Chem 270:11,962
64. Tanaka S, Saito K, Reed JC (1993) J Biol Chem 268:10,920
65. Yang E, Zha J, Jockel J, Boise LH, Thompson CB, Korsmeyer SJ (1995) Cell 80:285
66. Korsmeyer SJ, Shutter JR, Veis DJ, Merry DE, Oltvai ZN (1993) Sem Cancer Biol 4:327
67. Sedlak TW, Oltvai ZN, Yang E, Wang K, Boise LH, Thompson CB, Korsmeyer SJ (1995) Proc Natl Acad Sci USA 92:7834
68. Apte SS, Mattei MG, Olsen BR (1995) Genomics 26:592
69. Olsen CW, Kehren JC, Dybdahl- Sissoko NR, Hinshaw VS (1996) J Virol 70:663
70. Miyashita T, Reed JC (1995) Cell 80:293
71. Zhan Q, Fan S, Bae I, Guillouf C, Liebermann D, O'Connor P, Fornace A (1994) Oncogene 9:3743
72. Miyashita T, Krajewski S, Krajewska M, Wang HG, Lin HK, Liebermann DA, Hoffman B, Reed JC (1994) Oncogene 9:1799
73. Knudson CM, Tung KS, Tourtellotte WG, Brown GA, Korsmeyer SJ (1995) Science 270:96
74. Lotem J, Sachs L (1995) Cell Growth Diff 6:647
75. Miyashita T, Kitada S, Krajewski S, Horne WA, Delia D, Reed JC (1995) J Biol Chem 270:26,049
76. Zha H, Aime-Sempe C, Sato T, Reed JC (1996) J Biol Chem 271:7440
77. Simonian PL, Arillot DAM, Andrews DW, Leber B, Nunez G (1996) J Biol Chem 271:32,073
78. Krajewski S, Krajewska M, Shabaik A, Reed JC (1994) Am J Pathol 145:1323
79. Bargou RC, Daniel PT, Mapara MY, Bommer K, Wagner C, Kallinich B, Royer HD, Dorken B (1995) Int J Cancer 60:854
80. Krajewski S, Blomqvist C, Franssila K, Krajewska M, Wasenius VM, Niskanen E, Nordling S, Reed JC (1995) Cancer Res 55:4471
81. Yin C, Knudson CM, Korsmeyer SJ, Van Dyke T (1997) Nature 385:637
82. Rampino N, Yamamoto H, Ionov Y, Li Y, Sawai H, Reed JC, Perucho M (1997) Science 275:967
83. Boise LH, Gonzalez-Garcia M, Postema CE, Ding L, Lindsten T, Turka LA, Mao X, Nunez G, Thompson CB (1993) Cell 74:597
84. Sato T, Hanada M, Bodrug S, Irie S, Iwama N, Boise LH, Thompson CB, Golemis E, Fong L, Wang HG (1994) Proc Natl Acad Sci USA 91:9238
85. Gonzalez-Garcia M, Perez-Beallestro R, Ding L, Duan L, Boise LH, Thompson CB, Nunez G (1994) Development 120:3033
86. Rouayrenc JF, Boise LH, Thompson CB, Privat A, Patey GCR (1995) Acad Sci III 318:5537
87. Park JR, Bernstein ID, Hockenbery DM (1995) Blood 86:868
88. Motoyama N, Wang F, Roth KA, Sawa H (1995) Science 267:1506

89. Choi MS, Boise LH (1995) Eur J Immunol 25:1352
90. Gottschalk AR, Boise LH, Thompson CB, Quintans J (1994) Proc Natl Acad Sci USA 91:7350
91. Muchmore SW, Sattler M, Liang H, Meadows RP, Harlan JE, Yoon HS, Nettesheim D, Chang BS, Thompson CB, Wong S, Ng S, Fesik SW (1996) Nature 381:335
92. Sattler M, Liang H, Nettesheim D, Meadows RP, Harlan JE, Eberstadt M, Yoon HS, Shuker BS, Chang BS, Minn AJ, Thompson CB, Fesik SW (1997) Science 275:983
93. Chittenden T, Harrington EA, O'Connor R, Flemington C, Lutz RJ, Evan GI, Guild BC (1995) Nature 374:733
94. Kiefer MC, Brauer MJ, Powers VC, Wu JJ, Umansky SR, Tomel LD, Barr PJ (1995) Nature 374:736
95. Farrow SN, White JHM, Martinou I, Raven T, Pun K, Grinham CJ, Martinou J, Brown R (1995) Nature 374:731
96. Moss SF, Agarwal B, Arber N, Guan RJ, Krajewska M, Krajewski S, Reed JC Holt PR (1996) Biochem Biophys Res Comm 223:199
97. Krajewski M, Moss SF, Krajewski S, Song K, Holt PR, Reed JC (1996) Cancer Res 56:2422
98. Yang E, Zha J, Boise LH, Thompson CB, Korsmeyer SJ (1995) Cell 80:285
99. Zha J, Hanada H, Yang E, Jockel J, Korsmeyer SJ (1996) Cell 87:619
100. Wang H-G, Rapp UR, Reed JC (1996) Cell 87:629
101. Kozopas KM, Yang T, Buchan HL, Zhou P, Craig RW (1993) Proc Natl Acad Sci USA 90:3516
102. Yang T, Kozopas KM, Craig RW (1995) Cell Biol 128:1173
103. Craig RW, Jabs EW, Zhou P, Kozopas KM, Hawkins AL, Rochelle JM, Seldin MF, Griffin CA (1994) Genomics 23:457
104. Atkin N (1986) Cytogenetic 21:279
105. Gendler SJ, Cohen EP, Craston A, Duhig T, Johnstone G, Barnes D (1990) Int J Cancer 45:431
106. Testa JR (1990) Cell Growth Diff 1:97
107. Reynolds JE, Yang T, Qian L, Jenkinson JD, Zhou P, Eastman A, Craig RW (1994) Cancer Res 54:6348
108. Lomo J, Smeland EB, Krajewski S, Reed JC (1996) Cancer Res 56:40
109. Krajewski S, Bodrug SE, Krajewski M, Shabik A, Gascoyne R, Berean K, Reed JC (1995) Am J Path 146:1309
110. Krajewski S, Bodrug SE, Gascoyne R, Berean K, Krajewski M, Reed JC (1994) Am J Path 145:515
111. Lin EY, Orlofsky A, Berger MS, Prystowsky MB (1993) J Immunol 151:1979
112. Karsan A, Yee E, Harlan M (1996) J Biol Chem 271:27,201
113. Choi SS, Park IC, Yun JW, Sung YC, Hong SI, Shin HS (1995) Oncogene 11:1693
114. D'Sa-Eipper C, Subramanian T, Chinnadurai G (1996) Cancer Res 56:3879
115. Lucas JM, Bryans M, Lo K, Wilkie NM, Freshney M, Thornton D, Lang JC (1994) Oncol Res 6:139
116. Kenny JJ, Knobloch TJ, Augustus M, Carter KC, Rosen CA, Lang JC (1997) Oncogene 14:997
117. Wang K, Yin X, Chao DT, Milliman CL, Korsmeyer SJ (1996) Genes Dev 10:2859
118. Boyd JM, Gallo GJ, Elangovan B, Houghton AB, Malstrom S, Avery BJ, Ebb RG, Subramanian T, Chittenden T, Lutz RJ, Chinnadurai G (1995) Oncogene 11:1921
119. Orth K, Dixit VM (1997) J Biol Chem 272:8841
120. Gibson L, Holmgreen SP, Huang DCS, Bernard O, Copeland NG, Jenkins NA, Sutherland GR, Baker E, Adams JM, Cory S (1996) Oncogene 13:665
121. Inohara N, Ding L, Chen S, Nunez G (1997) EMBO J 16:1686
122. Cheng, EH-Y, Levine, B, Boise, LH, Thompson, CB, Hardwick, JM (1996) Nature 379:554
123. Reed JC, Zha H, Aime-Sempe C, Takayama S, Wang HG (1996) Adv Exp Med Biol 406:99
124. Shibasaki F, Kondo E, Akagi T, McKeon F (1997) Nature 386L:728
125. Chen C-Y, Faller DV (1996) J Biol Chem 271:2376
126. Wang H-G, Takayama S, Rapp UR, Reed JC (1996) Proc Natl Acad Sci USA 93:7063

127. Wang H-G, Millan JA, Cox AD, Der CJ, Rapp UR, Beck T, Zha H, Reed JC (1995) J Cell Biol 129:1103
128. Fernandez A, Marin MC, McDonnell TJ, Ananthaswamy HN (1994) Oncogene 9:2009
129. Jaattela M, Tewari M, Shayman JA, Dixit VM (1995) Oncogene 10:2297
130. Talley AK, Dewhurst S, Perry SW, Dollard SC, Gummuluru S, Fine SM, New D, Epstein LG, Gendelman HE, Gelbard HA (1995) Mol Cell Biol 15:2359
131. Vandenabeele P, Declercq W, Vanhaesebroeck B, Grooten J, Fiers W (1995) J Immunol 154:2904
132. Miura M, Zhu H, Rotello R, Hartwieg EA, Yuan J (1993) Cell 75:653
133. Hockenbery DM, Oltvai ZN, Yin XM, Milliman CL, Korsmeyer SJ (1993) Cell 75:241
134. Kane DJ, Sarafian TA, Anton R, Hahn H, Gralla EB, Valentine JS, Ord T, Bredesen DE (1993) Science 262:1274
135. Martin SJ, Newmeyer DD, Mathias S, Farschon DW, Wang HG, Reed JC, Kolesnick RN, Green DR (1995) EMBO J 14:5191
136. Strasser A, Anderson RL (1995) Cell Growth Differ. 6:799
137. Baffy G, Miyashita T, Williamson JR, Reed JC (1993) J Biol Chem 268:6511
138. Lam M, Dubyak G, Chen L, Nunez G, Miesfeld RL, Distelhorst CW (1994) Proc Natl Acad Sci USA 91:6569
139. Marin MC, Fernandez A, Bick RJ, Brisbay S, Buja LM, Snuggs M, McConkey DJ, von Eschenbach AC, Keating MJ, McDonnell TJ (1996) Oncogene 11:2259
140. Hsu B, Marin MC, Brisbay S, McConnell K, McDonnell TJ (1994) Cancer Bull 46:125
141. Miyashita T, Reed JC (1992) Cancer Res 52:5407
142. Raffo AJ, Perlman H, Chen MW, Day ML, Streitman JS, Buttyan R (1995) Cancer Res 55:4438
143. Batistatou A, Merry DE, Korsmeyer SJ, Greene LA (1993) J Neurosci 13:4422
144. Sentman CL, Shutter JR, Hockenbery D, Kanagawa O, Korsmeyer SJ (1991) Cell 67:879
145. Herrmann JL, Bruckheimer E, McDonnell TJ (1996) Biochem Soc Trans 24:1059
146. Verheij M, Bose R, Lin XH, Yao B, Jarvis WD, Grant S, Birrer MJ, Szabo E, Zon LI, Kyriakis JM, Haimovitz-Friedman A, Fuks Z, Kolesnick RN (1996) Nature 380:75
147. Liu ZG, Hsu H, Goeddel DV, Karin M (1996) Cell 87:565
148. Scutze S, Potthoff K, Machleidt T, Berkovic D, Weigmann K, Kronke M (1992) Cell 71:1
149. Lin K-L, Lee S-L, Narayanan R, Baraban JM, Hardwick JM, Ratan RR (1995) J Cell Biol 131:1149
150. Lee FS, Hagler J., Chen ZJ, Maniatis T (1997) Cell 88:213
151. Thanos D, Maniatis T (1995) Cell 80:529
152. Herrmann JL, Beham AW, Sarkiss M, Chiao PJ, Rands MT, Bruckheimer E, McDonnell TJ (1997) Exp Cell Res 237:101
153. Van Antwerp DJ, Martin SJ, Kafri T, Green DR, Verma IM (1996) Science 274:787
154. Wu M, Lee H, Bellas RE, Schauer SL, Arsura M, Katz D, Fitzgerald MJ, Rothstein TL, Sherr DH, Sonenshein GE (1996) EMBO J 15:4682
155. Akao Y, Otsuki Y, Kataoka S, Ito Y, Tsujimoto Y (1994) Cancer Res 54:2468
156. Krajewski S, Tanaka S, Takayama S, Schibler MJ, Fenton W, Reed JC (1993) Cancer Res 53:4701
157. Richter C, Gogvadze V, Laffranchi R, Schlapbach R, Schweizer M, Suter M, Walter P, Yafee M (1995) Biochim Biophys Acta 1271:67
158. Lotscher HR, Winterhalter KH, Carafoli E, Richter C (1980) J Biol Chem 255:9325
159. Frei B, Winterhalter KH, Richter C (1986) Biochemistry 25:4438
160. Hennet T, Richter C, Peterhans E (1993) Biochem J 289:587
161. Bajjaleih SM, Martin TFJ, Floor E (1989) J Biol Chem 264:14,354
162. Chatterjee S (1993) Adv Lipid Res 26:25
163. Ryan JJ, Prochownik E, Gottlieb CA, Apel IJ, Merino R, Nunez G, Clarke MF (1994) Proc Natl Acad Sci USA 91:5878
164. Meikrantz W, Gisselbrecht S, Tam SW, Schlegel R (1994) Proc Natl Acad Sci USA 91:3754
165. Kumar S, Lavin MF (1996) Cell Death Diff 3:255
166. Cohen GM (1997) Biochem J 326:1

167. Naumovski L, Cleary ML (1996) Mol Cell Biol 16:3884
168. Zamzami N, Susin SA, Marchett P, Hirsch T, Gomez-Monterrey I, Castedo M, Kroemer G (1996) J Exp Med 183:1533
169. Smets LA, Van der Berg J, Acton D, Top B, van Rooij H, Verwijs-Janssen M (1994) Blood 5:1613
170. Minn AJ, Velez P, Schendel SL, Liang H, Muchmore SW, Fesik SW, Fill M, Thompson CB (1997) Nature 385:353
171. Yang J, Liu X, Bhalla K, Kim CN, Ibrado AM, Cai J, Peng, Jones DP, Wang X (1997) Science 275:1129
172. Kluck RM, Bossy-Wetzel E, Green DR, Newmeyer DD (1997) Science 275:1132
173. Xia Z, Dickens M, Raingeaud J, Davis RJ, Greenberg ME (1995) Science 270:1326
174. Ito T, Deng X, Carr B, May WS (1997) J Biol Chem 272:11,671

Received January 1998

The Role of Caspases in Apoptosis

Natasha L. Harvey[1] · Sharad Kumar[2]

[1] The Hanson Centre for Cancer Research, Institute of Medical and Veterinary Science,
PO Box 14, Rundle Mall, Adelaide, South Australia, 5000
E-mail: nharvey@immuno.imvs.sa.gov.au
[2] The Hanson Centre for Cancer Research, Institute of Medical and Veterinary Science,
PO Box 14, Rundle Mall, Adelaide, South Australia, 5000,
E-mail: skumar@immuno.imvs.sa.gov.au

The process of apoptosis, or programmed cell death, is fundamental during normal development and homeostasis and aberrant apoptosis has been implicated in a number of human diseases. The cellular machinery involved in the execution of apoptosis includes a family of cysteine proteases termed caspases. Caspases exhibit the rare substrate preference of cleavage C-terminal to aspartate residues, a property shared only by the cytotoxic lymphocyte serine protease, granzyme B. Experimental evidence demonstrates a vital role for caspase activation in the apoptotic pathway, and, as such, caspases are a target for the development of agents that can modulate their activity. This article reviews the members of the caspase family and the role that each contributes to the execution of cell death induced by apoptotic stimuli.

Keywords: Cysteine protease, Caspase-1, Caspase-2, Caspase-3, Caspase-4, Caspase-5, Caspase-6, Caspase-7, Caspase-8, Caspase-9, Caspase-10, Caspase-11, Granzyme B, CED-3, CED-4, CED-9, Bcl-2, PARP, Fas/APO1, TNF.

Advances in Biochemical Engineering/
Biotechnology, Vol. 62
Managing Editor: Th. Scheper
© Springer-Verlag Berlin Heidelberg 1998

List of Symbols and Abbreviations

ICE   interleukin-1β-converting enzyme
PARP   poly(ADP-ribose)-polymerase
TNF   tumour necrosis factor

1
Introduction

Apoptosis is the term first given by Kerr and colleagues [1] to the active mode of programmed cell death that is distinct from necrosis and is characterised by a defined progression of morphological changes in the dying cell. The morphological characteristics of apoptosis include plasma membrane blebbing, nuclear condensation, loss of cell volume and fragmentation of DNA at nucleosomal intervals. In addition to these features, apoptosis is accompanied by the absence of inflammation, as opposed to necrosis which induces an inflammatory response [1,2]. The execution of apoptosis is fundamental during development and homeostasis and dysregulation of the apoptotic pathway has been implicated in a number of human diseases. Diseases associated with the failure of cells to undergo apoptosis include cancer and autoimmune disorders. Conversely, excessive apoptosis has been implicated in neurodegenerative disorders such as Alzheimer's disease, Parkinson's disease and spinal muscular atrophy. In addition, inappropriate apoptosis is associated with AIDS, ischaemic heart damage, viral infections and aplastic anaemia [3]. Elucidation of the mechanisms involved in the execution of apoptosis is thus an area of intense research. Emerging evidence demonstrates the role of a family of aspartate specific cysteine proteases recently termed caspases [4]. The proteolytic activation of caspases in response to apoptotic stimuli mediates the series of specific cellular events that occurs in the execution phase of apoptotic cell death. The development of agents that selectively interact with the apoptotic machinery to either inhibit inappropriate apoptosis or selectively initiate apoptosis in cells that are abnormally resistant will be crucial to combat the diseases in which aberrant apoptosis occurs.

2
Programmed Cell Death in *Caenorhabditis elegans* Development

Of 1090 cells generated during *C. elegans* development, 131 are removed by a programmed process resembling apoptosis [5, 6]. To date, 14 genes have been identified that are involved at various stages of the cell death pathway (reviewed in [7]). The products of these genes function in the initiation of apoptosis, the execution of apoptosis and the engulfment of apoptotic cells. Of the genes involved in the execution of cell death, two, *ced-3* and *ced-4*, are absolutely required for cell death to occur. Loss of function mutations in either of these genes results in extra cells in *C. elegans* embryos [8]. In contrast, the *ced-9* gene product is involved in the protection of cells against apoptosis; loss of fuction mutations in this gene result in the death of cells that normally survive [9]. Genetic studies have shown that *ced-9* fuctions upstream of *ced-3* and *ced-4* to negatively regulate their apoptosis inducing activity [9] and thus maintain cell survival. *ced-9* encodes a protein homologous to the product of the mammalian *bcl-2* proto-oncogene [10] which protects cells from apoptosis [11]. In fact, *bcl-2* expression can prevent apoptosis in *C. elegans* [12] and partially compensate for *ced-9* mutations [10], indicating that these molecules are functionally similar. These observations suggested that the molecules involved in the apoptotic pathway have been conserved throughout evolution.

3
ICE and Nedd2, the First Mammalian Homologues of CED-3

The cloning and characterisation of *ced-3* [13] and its protein product revealed homology with the human interleukin-1β-converting enzyme (ICE) and murine *Nedd2*. ICE is a cysteine protease required for cleavage of the inactive 31 kDa precursor of human interleukin-1β to its mature 17.5 kDa active form [14, 15], while *Nedd2* was identified as a *neural precursor cell expressed* gene that is *downregulated during development* of the murine brain [16]. *Nedd2* was subsequently shown to encode a cysteine protease homologous to ICE [17, 18]. Both ICE and *Nedd2* cleave C-terminal to an aspartate residue at the P_1 position [14, 15, 19, 20], a substrate specificity shared only by granzyme B, a serine protease found in the granules of cytotoxic lymphocytes that is required for their cytotoxic function (reviewed in [21]). Comparison of the *C. elegans* CED-3 protein sequence with related nematode species *C. briggsae* and *C. vulgaris* demonstrated most conservation in the carboxy terminal half of the protein, the region most homologous to ICE and Nedd2 [13]. The three nematode CED-3 proteins share 29% amino acid identity with ICE [13]; absolutely conserved residues including those that surround the catalytic cysteine of ICE, QACRG [15]. The homology shared by ICE and CED-3 suggested that CED-3 may exert its apoptosis inducing function in *C. elegans* by acting as a cysteine protease and also that ICE and Nedd2 may play an equivalent role to that of CED-3 in the execution of mammalian apoptosis. Indeed, ICE and Nedd2 were subsequently shown to possess apoptosis inducing activity [17, 18, 21], indicating that in mammals, at least two functional homologues of CED-3 exist that initiate the onset of apoptosis.

4
The Caspase Family

Subsequent to the description of the pro-apoptotic funtion of CED-3, ICE and Nedd2, a family of homologues have been identified, all of which are *c*ysteine prote*ases* that share the requirement for cleavage at *asp*artate residues. For uniformity, these proteases have recently been termed caspases [4] which derives from their function and substrate specificity. To date 11 caspases have been described and have been allotted numbers based on the chronological order of their discovery (Table 1). The caspases can be divided into three subfamilies based on sequence similarity and substrate specificity. The ICE subfamily comprises caspases-1, 4, 5 and 11, the CPP32 subfamily comprises caspases-3, 6, 7, 8, 9 and 10 and, to date, caspase-2 stands alone. In addition to this classification, caspases can be further divided on the basis of the presence or absence of an amino-terminal pro-domain, which is released upon protease activation. Emerging evidence suggests that caspases containing a long pro-domain are the first to be activated in response to apoptotic stimuli, while those with short or absent pro-domains are activated subsequently and are the "effector" proteases responsible for cleaving the cellular substrates that mediate the apoptotic death of the cell.

Table 1. Members of the caspase family

Caspase subfamily	Caspase members	Alternative names	References
Caspase-1	Caspase-1	ICE	14, 15
	Caspase-4	TX, ICH-2, ICE-rel-II	55–57
	Caspase-5	TY, ICE-rel-III	57, 58
	Caspase-11	ICH-3	59
Caspase-2	Caspase-2	Nedd2, ICH-1	16–18
Caspase-3	Caspase-3	CPP32, Yama, apopain	60–62
	Caspase-6	Mch2	67
	Caspase-7	Mch3, ICE-LAP3, CMH-1	68–70
	Caspase-8	MACH, FLICE, Mch5	81–83
	Caspase-9	ICE-LAP6, Mch6	84–87
	Caspase-10	Mch4, FLICE2	81, 87

5
The ICE Subfamily of Caspases

5.1
ICE as a Prototype for Caspase Structure and Function

Interleukin-1β (IL-1β) plays a major role in inflammation and as such is an attractive target for therapeutic agents that modulate its production and/or activity. The interleukin-1β-converting enzyme (ICE) (caspase-1) was originally identified in monocytes as a unique cysteine protease which is responsible for the cleavage of the 31 kDa interleukin-1β precursor molecule between Asp-116 and Ala-177 and Asp-27-Gly-28 to liberate the 17.5 kDa active cytokine [14, 15, 23, 24]. ICE encodes a 45 kDa protein which is proteolytically processed at aspartate residues to generate two subunits of 20 kDa and 10 kDa in size which heterodimerise to form the active enzyme [14, 15]. The catalytic cysteine residue of ICE is located in the p20 subunit although both subunits are required for enzyme activity [15]. The fact that the ICE pro-enzyme is cleaved at aspartate residues to liberate the p20 and p10 subunits suggested that autocatalysis may be the mechanism of ICE activation. This has been shown to be the case in vitro [15, 25, 26].

Based on the site at which ICE cleaves IL-1β, initial studies investigating the minimal recognition sequence of ICE have shown that there is a stringent requirement for aspartate adjacent to the cleavage site and also a requirement for a minimum of four amino acids amino-terminal to the cleavage site [27, 28]. Substitution studies performed by Thornberry and colleagues [15] demonstrated an increased affinity for the tetrapeptide sequence Ac-YVAD-NH-CH$_3$, which differs by one amino acid from the YVHD recognition sequence in pro-IL-1β. Thus far, YVAD has been the peptide substrate and inhibitor used to detect and inhibit ICE-like protease activity. Recent studies employing a positional scanning library have identified the optimal tetrapeptide sequence for ICE as WEHD [29], which may facilitate the identification of additional cellular substrates of ICE that are cleaved during apoptosis. Recently, a new substrate of caspase-1 has been identified as IFN-γ inducing factor, thus indicating a role for caspase-1 in the regulation of IFN-γ production [30, 31].

5.2
Crystal Structure of ICE

Derivation of the crystal structure of ICE complexed with the YVAD tetrapeptide demonstrated that the active enzyme is likely a tetramer comprised of two (p20 + p10) heterodimers [32, 33]. The spatial arrangement of the p20 and p10 subunits in the ICE tetramer highlights why both subunits are required to form the active site of the enzyme. There are two hypotheses as to how the p20 and p10 subunits associate to form the tetramer. The first possibility is that the p20 and p10 subunits of a heterodimer are derived from the same p45 precursor molecule and undergo significant structural rearrangment upon processing to form the confirmation in which they exist in the tetramer. The second possibility is that the p20 and p10 subunits comprising one heterodimer are derived from separate precursor molecules and associate once cleavage is complete. More recent evidence suggests that the pro-domain of ICE plays an important role in facilitating the correct subunit association to form active enzyme. Expression studies in yeast showed that p45 ICE was able to undergo autocatalytic processing while p30 ICE lacking the pro-domain was not [34]. In addition, p45 ICE in which the the active site cysteine was mutated retained the ability to dimerise [34], although processing was abolished. Dimerisation was shown to require absolutely the pro-domain and occur prior to autoprocessing, although the pro-domain was not sufficient for dimerisation and autoprocessing. It thus appears that the pro-domain of ICE plays a regulatory function in ICE autocatalysis, but whether autocatalysis is the primary mechanism of ICE activation in vivo or whether ICE activation requires the activity of another caspase member is presently not known.

5.3
The Role of Caspase-1 in Apoptosis

Caspase-1 (ICE) was the first homologue of CED-3 to be implicated in the apoptotic pathway in mammals when its overexpression in Rat-1 cells was shown to induce apoptosis [22]. The apoptosis induced by caspase-1 was inhibited by overexpression of *Bcl-2* or of the cowpox virus encoded gene *crmA* [22], first identified as a specific inhibitor of ICE [35]. Further evidence of a role for caspase-1 or caspase-1-like proteases came from Gagliardini and colleagues [36] who observed that CrmA could suppress apoptosis initiated by the withdrawal of NGF from neurons, suggesting that caspase-1 may participate in the execution of apoptosis in neural cells. CrmA has subsequently been shown to inhibit apoptosis in response to a variety of stimuli including serum deprivation [18], Fas/APO-1 [37–40], tumour necrosis factor [37,41], γ-irradiation [42] and cytotoxic T lymphocytes [37]. In addition, tetrapeptide inhibitors of ICE such as YVAD-CMK, YVAD-CHO and z-VAD-FMK have been shown to inhibit apoptosis induced by a variety of stimuli in a variety of cell types [38, 39, 43–46]. In retrospect, the inhibition of apoptosis observed in these experiments is probably not due solely to the inhibition of caspase-1 activity, but due to the ability of CrmA to inhibit other members of the caspase family such as caspase-8 [47]. Another observation suggesting a role for caspase-1 in the apoptotic pathway is the upregulation of caspase-1 expression in the apoptotic process of mouse mammary tissue involution [48].

5.4
The Role of Caspase-1 in Fas/APO-1 Mediated Apoptosis

The generation of caspase-1 null mice has shed some light on the role of caspase-1 in the apoptotic pathway. Caspase-1 null mice display an apparently normal phenotype, but are deficient in the production of IL-1α and IL-1β [49, 50]. Their apparently normal passage through development suggests that caspase-1 is not the functional mammalian equivalent of *ced-3*. The only apoptotic aberration in these mice was observed in thymocytes induced to undergo apoptosis by Fas/APO-1. While caspase-1 (-/-)thymocytes retained the ability to apoptose in response to dexamethasone and ionising radiation, they were resistant to apoptosis induced by Fas/APO-1 [49]. Abrogation of the Fas/APO-1 pathway in these cells was not complete as is the case in *lpr* and *gld* mice which have mutated Fas/APO-1 or Fas/APO-1 ligand respectively (reviewed in [51]). These mice display lymphadenopathy and splenomegaly due to a failure of T lymphocytes to undergo apoptosis and some develop autoimmune diseases (reviewed in [51]). These observations suggest that other caspases in addition to caspase-1 are involved in the Fas/APO-1 induced apoptotic pathway or that other caspase family members are able to compensate for the absence of caspase-1 due to functional redundancy. Additional evidence to support the role of ICE in Fas/APO-1 mediated apoptosis includes the measurement of caspase-1 protease activity on the YVAD-containing fluorogenic substrate following Fas/APO-1 ligation and the inhibition of Fas/APO-1 induced apoptosis by expression of antisense ICE [39]. Investigation of the activation of different subfamilies of caspases in response to Fas/APO-1 ligation revealed that caspase-1 or a caspase-1-like protease was activated transiently, while the activity of caspase-3 and/or caspase-3 family members gradually increased during Fas/APO-1 induced apoptosis [52]. The activation of caspase-3-like function was shown to be dependent on prior caspase-1-like activation. A similar observation has been made in vivo in response to administration of anti-Fas/APO-1 antibody which results in liver damage [53]. Rodriguez et al. [54] demonstrated the sequential activation of caspase-1-like activity followed by caspase-3-like activity as liver damage progressed and showed that both caspase activation and liver damage could be prevented by intravenous administration of z-VAD-FMK.

5.5
Caspase-4, Caspase-5 and Caspase-11

Caspase-4, caspase-5 and caspase-11 were identified based on sequence similarity with caspase-1 and have all been shown to induce apoptosis when overexpressed [55–59]. Caspases-1, 4 and 5 share just over 50% sequence identity with one another and 26–28% sequence identity with ced-3. Caspase-11 shares 45% sequence identity with caspase-1, 60% with caspase-4 and 54% with caspase-5. Neither caspase-4, caspase-5 nor caspase-11 are able to cleave pro-IL-1β [55–57, 59] and the substrate specificity of caspase-4 appears to diverge significantly from caspase-1 based on substrate and inhibitor binding studies [56]. While the K_m of caspase-1 for acetyl-YVAD-*p*-nitroanilide is 83 ± 10 µmol/l, the K_m of caspase-4 for the same substrate was found to be 681 ± 84 µmol/l [56]. Correspondingly, the IC_{50} of caspase-4 for YVAD-CHO is approximately 20 times higher (748 nmol/l) than that for caspase-1 (38 nmol/l), indicating divergent substrate preference and function of these family members [56]. Peptide cleavage studies performed by Talanian and colleagues [20] demonstrate that caspase-4 prefers the tetrapeptide substrate LEVD, with an aliphatic P_4 residue as opposed to the aromatic P_4 residue (Y) of caspase-1. Caspase-4 is able to cleave caspase-1 in vitro [55], and, similarly, caspase-11 promotes IL-1β processing by caspase-1 in vitro and in vivo [59], indicating that both caspase-4 and caspase-11 may function upstream of caspase-1 to initiate its activation.

6
The CED-3/Caspase-3 Subfamily

The caspase-3-like subfamily is composed of caspase-3 and its closest homologues caspases-6, 7, 8, 9 and 10. Caspase-3 was first cloned by Fernandes-Alnemri et al. [60] and shown to encode a 32 kDa cysteine protease homologous to CED-3, caspase-1 and caspase-2 that was capable of inducing apoptosis when overexpressed in insect cells. By sequence comparison with ICE, the active enzyme was thought to consist of subunits 20 kDa and 11 kDa in size, the only major difference between the sequences being the absence of a long, amino-terminal pro-domain in caspase-3. Caspase-3 was subsequently identified as the protease responsible for cleaving poly(ADP-ribose) polymerase (PARP) [61, 62]. PARP is a nuclear enzyme involved in DNA repair and was one of the first cellular substrates shown to be cleaved in apoptotic cells [63]. The cleavage of PARP from a 116 kDa protein to an 85 kDa fragment first described by Kaufmann [63] is one of the most commonly used hallmarks of apoptotic cells. Prior to the cloning of caspase-3, cleavage studies performed by Lazebnik et al. [43], demonstrated that PARP was cleaved C-terminal to an aspartate residue, thus implicating the activity of a caspase, and although the protease responsible for PARP cleavage resembled caspase-1 in its profile of inhibitors, purified caspase-1 was not able to cleave PARP in vitro. Nicholson and colleagues [62] demonstrated by electrospray mass spectroscopy analysis of the purified active enzyme that the subunits comprising active caspase-3 are 17 kDa and 12 kDa in size. The substrate specificity of caspase-3 was determined from the DEVD sequence at which PARP is cleaved and, correspondingly, the peptide inhibitor Ac-DEVD-CHO was shown to be a potent inhibitor of PARP proteolysis. Comparative studies showed that the ICE specific inhibitors Ac-YVAD-CHO and CrmA were much less potent inhibitors of caspase-3 activity than Ac-DEVD-CHO [62].

6.1
Crystal Structure of Caspase-3

The determination of the crystal structure of caspase-3 complexed with a specific tetrapeptide inhibitor revealed homology to caspase-1 structure and also highlighted differences responsible for the divergence in substrate specificity of these two proteases [64, 65]. As with caspase-1, proteolytically active caspase-3 is a tetramer composed of two large and two small subunits, which are 17 kDa and 12 kDa in the case of caspase-3 [62]. The most marked difference in the substrate binding region of the enzymes was at the P_4 position, consistent with the caspase-3 preference for aspartate in this position as opposed to the caspase-1 preference for tyrosine or hydrophobic residues. The enzyme subsite that binds the P_4 substrate residue in caspase-3 differs significantly in both spatial and chemical composition from that of caspase-1, being a compact pocket that closely surrounds the P_4 aspartate side chain as distinct from the large, shallow region in caspase-1 that binds the more extensive tyrosine side chain at P_4. Other residues important for caspase-3 substrate specificity that are conserved between at least some members of the caspase-3 subfamily include a tryptophan residue at position 348 and an insert of ten amino acids at position 381, which contribute importantly to the P_4 binding site of caspase-3 [64]. The regions of caspase-3 and caspase-1 that bind the P_1-P_3 residues of the substrate share much more similarity, with most conservation at the P_1 binding position, which probably accounts for the absolute requirement of the caspases for aspartate at P_1.

Use of the two peptide inhibitors YVAD-CHO and DEVD-CHO reinforces the difference in affinity that caspase-1 and caspase-3 have for their substrates. YVAD-CHO has a K_i for caspase-1 of 0.76 nmol/l [15] and an K_i of 12 µmol/l for caspase-3 [62]. Correspondingly, DEVD-CHO has a K_i for caspase-3 of 0.2 nmol/l [20, 62] and a K_i for caspase-1 of 17 nmol/l [62]. Studies investigating the substrate preferences of different members of the caspase family will improve our knowledge of the contribution that different caspases make to the apoptotic pathway as well as enable the identification of additional apoptotic substrates and novel peptide inhibitors specific for individual caspases.

6.2
Caspase-3 as a Functional Homologue of CED-3

In addition to sharing the highest sequence similarity with CED-3, caspase-3 appears to have the most functional homology with CED-3. Caspase-3 null mice die at 1–3 weeks of age and their brain development is profoundly disturbed due to decreased apoptosis [66]. This suggests that caspase-3 plays a critical role in the execution of apoptosis during development, at least in cells of the central nervous system. Thymocytes from these mice retain the ability to undergo apoptosis in response to a variety of stimuli, suggesting that other caspase-3-like proteases may compensate for caspase-3 function in other tissues.

6.3
Discovery of Caspase-3 Homologues

Caspase-6 and caspase-7 were originally identified by Fernandes-Alnemri et al. [67, 68] using a PCR-based approach and were shown to be more homologous to caspase-3 than to caspase-1. Caspase-7 was also identified independently by Duan et al. [69] and Lippke et al. [70]. Caspase-6 is a 293 amino acid caspase that, like caspase-3, lacks a long amino terminal pro-domain. Caspase-6 cleaves PARP to the identical sized fragments produced by caspase-3 and observed in apoptotic cells, indicating that it may fulfill some of the same functions as caspase-3 in apoptosis. A novel property of caspase-6 is its ability to cleave lamins to the sizes apparent in apoptotic cells [67], an observation implied earlier by Lazebnik et al. [71] who showed that cleavage of nuclear lamins during apoptosis in a cell free system required the activity of a caspase distinct from the one that cleaves PARP.

Orth and colleagues [72] also demonstrated that caspase-6 was responsible for lamin cleavage during apoptosis and showed that lamin cleavage activity was restricted to caspase-6 and was not mediated by caspase-3 or caspase-7. These results indicate that, although caspase-6 is closely related to caspase-3, it possesses unique substrate specificity.

Caspase-7 has proven to be a much closer homologue of caspase-3 than caspase-6 is, sharing 52 % amino acid identity with caspase-3 and 31 % identity with caspase-6 [70]. Caspase-7 is a 303 amino acid protein with a very short amino terminal pro-domain [68–70]. Interestingly, two forms of caspase-7 were identified by Fernandes-Alnemri et al. [68] and shown to be the result of alternative splicing; caspase-7 α is the apoptotic form while the alternatively spliced β form encodes a truncated protein that does not contain the QACRG active site and its function remains unknown [68]. Like caspases-3 and 6, caspase-7 can cleave PARP [68, 70], indicating possible functional similarity with caspase-3. Fernandes-Alnemri and colleagues [68] also made the interesting observation that the large and small subunits of caspase-3 and caspase-7 can associate together (caspase-3 p17 + caspase-7 p12 and vice versa) to form active enzyme and induce apoptosis in insect cells. Although this association has not been observed in vivo, it suggests the possibility of regulation of caspase activity by heterodimerisation with homologous caspase family members. The first description of subcellular localisation of a caspase family member was made with caspase-7, which was shown to diffusely spread throughout the cytoplasm [69]. More importantly, the first direct evidence of caspase activation in response to apoptotic stimuli in vivo was that of caspase-7 cleavage from the 35 kDa pro-enzyme to subunits 20 kDa and 12 kDa in size in response to treatment with Fas/APO-1 and tumour necrosis factor (TNF) [69]. Subsequently, caspase-3 and caspase-6 have also been shown to be activated in vivo in response to a variety of apoptotic stimuli [72–80]. Recent studies suggest that caspase-3 and caspase-6 are the major active caspases in apoptotic cells and, interestingly, that the species of active caspase subunits varies between cell lines, indicating differences in caspase processing [76, 79]. These results confirm that caspase activation is a vital process in the execution of cell death.

6.4
Caspases-8, 9 and 10

Although caspases-8, 9 and 10 belong to the caspase-3 subfamily on the basis of sequence similarity and substrate specificity, they differ by virtue of the presence of a long amino terminal pro-domain [81–85]. Caspase-8 was the first of these caspases to be identified and was isolated based on its physical interaction with the Fas/APO-1 associated adaptor molecule FADD/MORT1 [82, 83]. The physical coupling of the two proteins was shown to be mediated by the death effector domain of FADD/MORT1, binding analogous domains contained in the amino terminal pro-domain of caspase-8. This interaction serves to recruit caspase-8 to the Fas/APO-1 signalling complex in response to receptor ligation and presumably results in the proteolytic activation of caspase-8, directly linking caspase activation to the activation of a death receptor. Proteolytically active caspase-8 could then proceed to cleave cellular substrates that mediate the features of apoptotic death, including the activation of other caspase family members [85]. The association of FADD/MORT1 with TRADD, an analogous adaptor molecule that binds to TNFR-1 [86], therefore implicates caspase-8 recruitment and activation in apoptosis initiated by TNFR1 as well as by Fas/APO-1. Caspase-10 has a high level of sequence similarity with caspase-8, displaying 52% identity in their amino acid composition excluding the pro-domain [81]. In a similar manner to caspase-8, the pro-domain of caspase-10 contains FADD/MORT1 homologous death effector domains [81, 87], that bind the corresponding domain in FADD [87]. This serves to recruit caspase-10 to both Fas/APO-1 and TNF receptors in a FADD-dependent manner [87]. An important role for caspase-8 and caspase-10 in death signalling through these receptors is suggested by the fact that active site mutants of these caspases inhibit both Fas/APO-1 and TNF mediated apoptosis [87]. Both caspase-8 and caspase-10 have been shown to activate several other caspase members in vitro [85, 88], suggesting that their main function, once activated in vivo, may be to mediate this function, especially with regard to activating the other caspase-3-like proteases that possess short or absent pro-domains. It is interesting to note that caspases-8 and 10 are the first observed that deviate from the otherwise absolutely conserved QACRG sequence surrounding the catalytic cysteine, they contain the QACQG pentapeptide sequence [81]. Caspase-9 also possesses a deviation of the active site pentapeptide from QACRG to QACGG [84, 85]. Like caspases-8 and 10, caspase-9 possesses a long amino terminal pro-domain, although the function of this pro-domain has not been explored to date. As with all members of the caspase-3 subfamily, the overexpression of caspase-9 induces apoptosis and active caspase-9 has the ability to cleave PARP in vitro [84].

7
Caspase-2

Caspase-2 was originally identified as a *n*eural precursor cell *e*xpressed, *d*evelopmentally *d*ownregulated gene in the murine central nervous system [16] and was subsequently shown to encode a protein with homology to ced-3 and caspase-1 [17, 18]. To date, caspase-2 stands alone in the third subfamily based on sequence homology with other caspases and substrate preference. Caspase-2 shares 29% sequence identity with caspase-1 and 31% identity with CED-3. Several lines of evidence suggest a role for caspase-2 in apoptosis. Overexpression of caspase-2 induces apoptosis in a variety of cell types [17, 18, 89], while expression of antisense caspase-2 in factor-dependent cells delays the initiation of apoptosis induced by factor withdrawal [90, 91]. The expression of antisense caspase-2 in PC12 cells did not, however, protect against apoptosis induced by superoxide dismutase (SOD1) downregulation [91], suggesting that caspase-2 activity may be confined to selected apoptotic pathways. In an analogous manner to antisense experiments, an alternatively spliced form of caspase-2 that encodes a truncated protein has been shown to protect against cell death induced by serum withdrawal in Rat-1 and NIH-3T3 cells [18, 92]. Upregulation of caspase-2 mRNA has been observed in response to ischaemia-induced cell death in the Mongolian gerbil and in rat brain [93, 94], while downregulation of caspase-2 has been observed during gonadotropin-promoted follicular survival [95]. Like caspases-8 and 10, the amino terminal pro-domain of caspase-2 has recently been shown to mediate association with RAIDD, a death adaptor molecule involved in apoptosis induced by TNF [96, 97]. The association of caspase-2 and RAIDD with the TNF receptor complex is facilitated via coupling with the RIP and TRADD death adaptor molecules. The recently identified death receptor DR3 is more closely related to TNFR1 than to Fas/APO-1 and has been shown to bind TRADD and then recruit RIP and FADD molecules [98, 99]. It is thus feasible that caspase-2 may be recruited to the DR3 receptor as well as to the TNFR1 in response to apoptotic signals and thereby initiate the cascade of proteolysis that mediates apoptotic cell death. Caspase-2 has recently been shown to undergo rapid activation in response to a variety of apoptotic signals in various cell lines that are sensitive to apoptosis [78, 80] but not in cells that are resistant to apoptosis [80]. The activation of caspase-2 in response to apoptotic signals other than TNF indicates that caspase-2 may be recruited to other, as yet unidentified signalling complexes.

Although no cellular substrates of caspase-2 have yet been identified, studies with peptide substrates indicate that caspase-2 has a preference for a residue in the P_5 position and is only weakly active on a tetrapeptide substrate [20]. Caspase-2 prefers the sequence VDVAD, with a strong preference for aspartate in the P_4 position similar to caspase-3-like proteases . Correspondingly, the peptide inhibitor Ac-VDVAD-CHO was shown to inhibit caspase-2 with a K_i of 3.5 nmol/l as opposed to Ac-DEVD-CHO which inhibits caspase-2 at a much higher concentration (K_i 1.7 µmol/l) [20].

8
Substrates of Caspases in Apoptosis

Although the caspase-mediated cleavage of PARP is one of the hallmarks of apoptotic cell death, it is not sufficient to mediate the cell death process as mice deficient in PARP develop normally [100]. Many other cellular substrates that undergo cleavage during apoptosis have since been described. These include nuclear proteins, cytoskeletal proteins, cytoplasmic proteins and members of the caspase family as previously mentioned. One of the nuclear proteins degraded during apoptosis is the U1 small ribonucleoprotein (U1–70 kdal) which is involved in the production of mature mRNA and has been shown to be cleaved in response to UV irradiation, nutrient deprivation, Fas/APO-1 and TNF-induced apoptosis [101, 102]. The proteolysis of U1–70 kdal was shown not to be mediated by ICE [101] but was inhibited by CrmA [102] and has subsequently been shown to be cleaved by caspase-3 [103]. A second nuclear protein shown to be cleaved during apoptosis is the catalytic subunit of the DNA repair enzyme, DNA-dependent protein kinase (DNA-PK) [103–106]. DNA-PK is cleaved in a variety of cell types in response to various stimuli and studies by Song and colleagues have identified that caspase-3, but not caspases-1, 4 or 6, is capable of cleaving DNA-PKcs into the identical fragments observed in apopototic cells [104]. Similar results have been observed by Casciola-Rosen and colleagues

[103] and McConnell et al. [106]. The degradation of nuclear enzymes involved in DNA repair that results in their loss of function may facilitate the endonucleosomal cleavage of DNA that is characteristic of apoptotic cells. Other nuclear proteins cleaved by caspases during apoptosis include the sterol regulatory element binding proteins 1 and 2 which are involved in regulating cholesterol biosythesis [107–109] and the nuclear lamins A, C [71, 110] and B1 [63, 71, 111]. The cleavage of lamin A has been demonstrated specifically by caspase-6 but not caspase-3 or caspase-7 [72].

Cytoskeletal proteins cleaved by the action of caspases during apoptosis include actin, [112, 113], fodrin [114–116] and Gas2, a microfilament protein [117]. The cleavage of cytoskeletal proteins during apoptosis probably facilitates the morphological features of apoptosis such as membrane blebbing seen in dying cells. Cytoplasmic substrates cleaved during apoptosis include protein kinase Cδ [118], D4-GDI; a GDP *dissociation inhibitor* for the Rho family GTPases [119] and PITSLRE kinases which are a family of Cdc-2-like kinases thought to be involved in apoptotic signalling and tumourigenesis [120]. Additional caspase substrates cleaved during apoptosis include the tumour suppressor gene product Rb [121–123], hn RNP C1 and C2 [124], the large subunit of replication factor C (RFC140) [125] and the cytosolic phospholipase A2 [126]. The significance of proteolysis of all of these substrates is not clear.

8.1
The DNA Fragmentation Factor

Perhaps the crucial substrate of activated caspases is the most recently identified; DNA Fragmentation Factor (DFF) [127]. DFF was purified from the cytoplasm of HeLa cells as the protein that induces DNA fragmentation in isolated nuclei following activation by caspase-3. DFF is a novel, heterodimeric protein that is composed of two subunits 40 kDa and 45 kDa in size. Caspase-3 is responsible for cleavage of the 45 kDa subunit at two sites, thus generating the active factor capable of DNA fragmentation. Following caspase-3 mediated cleavage, there is no further requirement for caspase-3 or any other cytoplasmic factors. These observations directly link the activation of caspase-3 with the activation of DFF, that in turn is responsible for mediating DNA fragmentation by an as yet unknown mechanism.

9
Granzyme B Functions by Activating Caspases

Cytotoxic lymphocytes mediate the death of their target cells by inducing apoptosis, a process which requires the action of serine proteases termed granzymes contained in their cytotoxic granules (reviewed in [21, 128]). Of the granzymes contained in cytotoxic lymphocytes, granzyme B is the most abundant and cleaves C-terminal to aspartate residues, a specificity shared only by caspases. This common substrate specificity led to the suggestion that granzyme B may induce apoptosis in target cells by activating members of the caspase family [129], and this has been shown to be the case. Granzyme B is able to activate caspases-2, 3, 7, 8, 9, 10 and 11 [59, 81, 84, 130, 131], but not caspase-1 [132]. The sites cleaved in caspases-3, 7 and 10 demonstrate that granzyme B cleaves these proteins at an IXXD sequence to release the subunits which comprise the active protease [81]. Inhibition of caspase-3-like activity has been shown to partially prevent granzyme B induced apoptosis [133]. The ability of granzyme B to cleave and activate caspases potentiates an extremely efficient mechanism of initiating the death program in cells targeted by cytotoxic lymphocytes.

10
Inhibition of Caspase Function by p35

While CrmA inhibition of caspases is more restricted to caspase-1-like proteases and caspase-8, the baculoviral p35 inhibitor of apoptosis is more general, and has the ability to block the activity of several caspases with varying substrate specificities [19, 42, 134]. p35 inhibits the apoptosis of baculovirus infected insect cells [135] and is able to prevent the developmental apoptosis that occurs in *C. elegans* [136]. p35 has also been shown to inhibit the apoptosis of mammalian neural cells induced to undergo apoptosis by factor deprivation [137, 138], of NIH-3T3 cells deprived of serum [42], of apoptosis induced by γ-irradiation [42] and by Fas/APO-1 or TNF [139]. The mechanism of p35 inhibition was shown to reside in complex formation between the respective caspase and p35 due to the recognition of p35 as a substrate [134]. This was confirmed by Xue and Horvitz [19] and Bertin et al. [140], who showed that cleavage of p35 between Asp-87 and Gly-88 was absolutely required, but not sufficient, for its inhibitory action. Recently, another class of virally encoded anti-apoptotic proteins has been discovered and termed v-FLIPS (*viral FLICE-inhibitory proteins*). These proteins have homology with the death effector domains of FADD and caspase-8 [141, 142] and mediate inhibition of apoptosis by binding to FADD and preventing subsequent caspase-8 recruitment [142], acting to prevent the initiation of the apoptotic cascade.

11
Regulation of Caspase Activation

The activation of caspases in response to apoptotic stimuli is now well documented as an integral component of the cell death pathway. As such, the initial activation of these proteases is likely to be under stringent control so that inappropriate activation leading to premature cell death does not occur. In an analogous manner to the protective action of CED-9 upstream of CED-3 and CED-4 in the *C. elegans* pathway, Bcl-2 has been shown to act upstream of caspases to prevent their activation in the mammalian apoptotic pathway [143–147]. The mechanism of Bcl-2 inhibition of caspase activity appears indirect; Bcl-2 cannot inhibit caspase function post caspase activation [146]. While CED-9 appears to be the sole protein of its kind in *C. elegans*, there exists a family of Bcl-2 homologues in mammals (reviewed in [148]). Some of these, including Bcl-2, Bcl-x$_L$,

Mcl-1 and Bcl-w promote cell survival, while others such as Bax, Bcl-x$_S$, Bik, Bad and Bak promote cell death. All members of the Bcl-2 family share conserved domains that facilitate the formation of homo- and hetero-dimers which can regulate their activity. As an example, Bax is able to associate with Bcl-2 in vitro and in vivo and, by virtue of this assocation, suppresses the ability of Bcl-2 to inhibit apoptosis [149]. The ratio of anti-apoptotic to pro-apoptotic Bcl-2 family members is therefore one stage at which the decision between apoptosis induction or protection can be made (reviewed in [150]).

Recent evidence demonstrates how the central three apoptosis genes of *C. elegans* CED-3, CED-4 and CED-9 may interact together at the molecular level to control and mediate apoptosis [151–153]. Spector and colleagues demonstrated using the yeast two-hybrid assay that CED-9 physically interacts with CED-4. Previously characterised mutations that inhibit CED-9 activity were shown to disrupt CED-9 interaction with CED-4, suggesting that CED-9/CED-4 binding is important for the cell death protective function of CED-9 [151]. A more extensive study by Chinnaiyan et al. [152] showed that CED-4 induced apoptosis in mammalian cells could be inhibited by CED-9 and also by Bcl-x$_L$ via direct physical association between CED-4 and CED-9 or Bcl-x$_L$. CED-4 induced apoptosis was also shown to be inhibited by the caspase inhibitors p35, CrmA and zVAD-FMK as well as by an active site mutant of CED-3, implying that CED-4 induced apoptosis is reliant on the activation of CED-3. Following this, it was shown that CED-4 could bind CED-3, caspase-1 or caspase-8, but not caspases-3 and 6 which contain only a short pro-domain. Lastly, CED-4 was shown to bind CED-9 and CED-3 simultaneously and was absolutely required for this trimeric association to occur [152]. This suggests a model of cell death regulation whereby CED-4 and its not yet described human equivalent/s act as a link between initiating apoptosis by allowing CED-3/caspase activation and inhibiting apoptosis via physical association with CED-9/Bcl-2 family members (Fig. 1).

Wu and colleagues [153] confirmed the binding of CED-4 with CED-9 and demonstrated that co-expression of these two proteins in mammalian cells resulted in the relocation of CED-4 from a diffuse cytoplasmic distribution to the compact granular pattern associated with the mitochondrial and perinuclear location of CED-9 and Bcl-2. By extrapolation from the results of Chinnaiyan and colleagues [152] showing simultaneous CED-9/CED-4/CED-3 association, it is possible that CED-3 and caspase family members may also be located at mitochondrial and perinuclear membranes in complexes with CED-9/Bcl-2 and CED-4/mammalian equivalent. These studies also elegantly demonstrate that the primary function of Bcl-2/Bcl-xL may be to sequester a CED-4 homologue together with caspases, thereby preventing caspase autoactivation. The mechanism by which these complexes are disrupted to facilitate CED-4 induced activation of CED-3 and execution of apoptosis is currently not known.

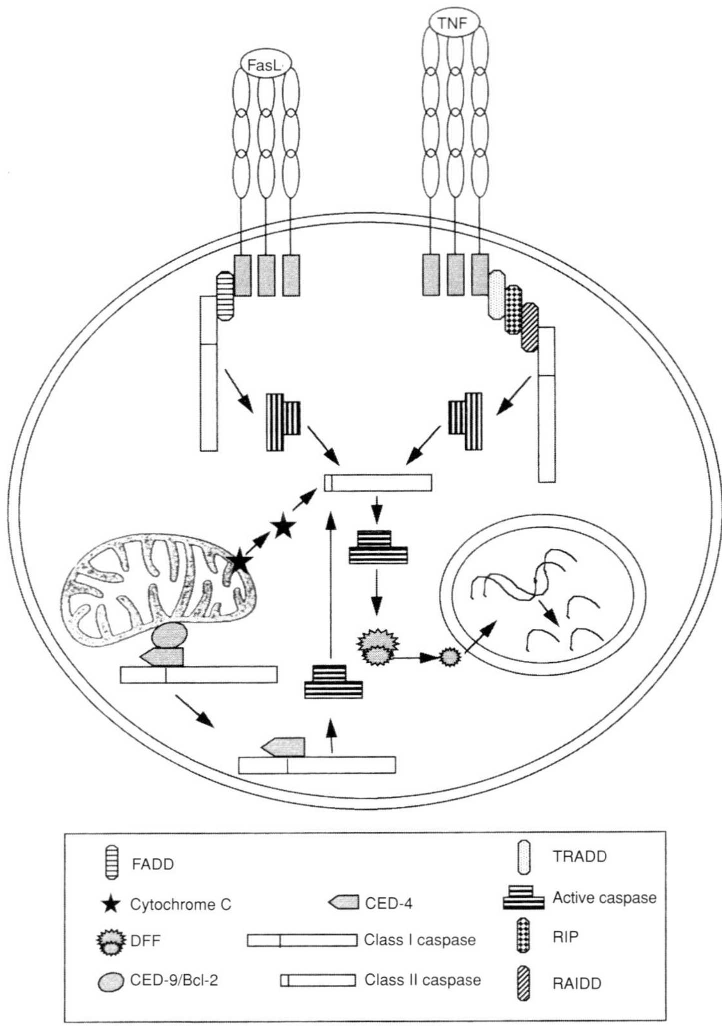

Fig. 1. A model for caspase activation. This model demonstrates the recruitment of caspases containing long pro-domains (Class I caspases) to death signalling complexes via adaptor molecules such as FADD (associated with the Fas/APO-1 receptor), TRADD, RIP and RAIDD (associated with recruitment of caspase-2 to TNF-R1). The activation of Class I caspases such as caspase 2, 8, 9 and 10 probably facilitates the activation of effector caspases (Class II caspases) such as caspase-3, 6 and 7 which have short pro-domains. Activation of these caspases subsequently mediates the cleavage of a variety of cellular substrates including DFF, which effects DNA fragmentation. Inhibition of caspase function can be mediated by CED-9 and its mammalian homologue Bcl-2, located at the mitochondrial membrane. These proteins may hold caspases in an inactive state via forming a trimeric complex with CED-4 and its putative mammalian homologue, until an appropriate death stimulus is received. This presumably releases CED-4 and CED-3/caspases, enabling the caspases to undergo proteolytic activation. Cytochrome C has also been shown to be released from mitochondria following apoptotic stimuli and serves to activate caspases by an as yet unknown mechanism

12
The Role of Cytochrome c in Caspase Activation

Cytochrome c is a mitochondrial protein located in the intermembrane space that is involved in the respiratory chain and has recently been implicated in the apoptotic program. Cytochrome c has been shown to be released to the cytosol in cells undergoing apoptosis [154–156], a process which is inhibited by Bcl-2 [155, 156]. Yang and colleagues [155] show that Cytochrome c release from mitochondria to the cytosol in response to staurosporine and etoposide induced apoptosis is blocked by expression of Bcl-2 or Bcl-x$_L$, which also inhibits caspase-3 activation, PARP cleavage and DNA fragmentation. In vitro experiments demonstrated that caspase-3 cleavage was absolutely dependent on Cytochrome c. Kluck et al. [156] found similar results and, in addition, showed that the addition of exogenous Bcl-2 to cytosol containing exogenous Cytochrome c was not able to inhibit any of the apoptotic effects induced by this cytosol. This observation suggests that Bcl-2 acts upstream of Cytochrome c to prevent its release from mitochondria and further experiments determined that Bcl-2 acts directly on the mitochondria, not on another cytoplasmic event responsible for Cytochrome c release. The direct mechanisms of Bcl-2 inhibition of Cytochrome c release from mitochondria and subsequent Cytochrome c mediated caspase activation are presently not known. These results suggest that Bcl-2 prevents apoptosis not only by sequestering CED-4 and CED-3 homologues at the mitochondrial membrane, but also by preventing the mitochondrial release of Cytochrome c.

13
Future Prospects

The progress in characterising the molecular mechanisms of apoptosis in the next few years will no doubt mirror the remarkable achievements since the cloning of CED-3 in 1993. There are many vital issues to be addressed, the most immediate of which includes identification of a mammalian CED-4 homologue, exploration for new caspase family members, further elucidation of the mechanism of death signal transmission resulting in caspase activation and investigation of CED-9/CED-4/CED-3 interaction. Although evidence confirms that caspases play an integral role in the execution of apoptosis, there exists the possibility that caspases perform additional, non-apoptotic roles in an analogous manner to the processing of IL-1β and regulation of IFN-γ production by caspase-1. Current investigations are no doubt exploring these and other imminent questions, the answers to which will further elucidate the role that caspases play in apoptosis.

Recently, a human homologue of CED-4 has been identified and termed Apaf-1 [157]. Apaf-1 is a 130 kDa protein comprised of an amino terminal CED-3 homologous domain, followed by a CED-4 homologous domain and carboxy terminal WD-40 repeats which are thought to mediate protein-protein interaction. It is thought that Apaf-1 may function to activate caspase-3 in an ATP-dependent manner following its association with cytochrome c and Apaf-3.

Whether mammalian cells contain a family of CED-4 homologues analogous to the family of CED-3 homologues is no doubt under current investigation.

14
References

1. Kerr JFR, Wyllie AH, Currie AR (1972) Br J Cancer 26:239
2. Wyllie AH, Kerr, JFR, Currie AR (1980) Int Rev Cytol 68:251
3. Thompson CB (1995) Science 267:1456
4. Alnemri ES, Livingston DJ, Nicholson DW, Salvesen G, Thornberry NA, Wong WW, Yuan J (1996) Cell 87:171
5. Sulston JE, Horvitz HR (1977) Dev Biol 82:110
6. Sulston JE, Schierenberg E, White JG, Thomson N (1983) Dev Biol 100:64
7. Horvitz HR, Shaham S, Hengartner MO (1994) Cold Spring Harbor Symp Quant Biol 59:377
8. Ellis HM, Horvitz HR (1986) Cell 44:817
9. Hengartner MO, Ellis RE, Horvitz HR (1992) Nature 356:494
10. Hengartner MO, Horvitz HR (1994) Cell 76:665
11. Vaux DL, Cory S, Adams JM (1988) Nature 335:440
12. Vaux DL, Weissman IL, Kim SK (1992) Science 258:1955
13. Yuan J, Shaham S, Ledoux S, Ellis HM, Horvitz HR (1993) Cell 75:641
14. Cerretti DP, Koslosky CJ, Mosley B, Nelson N, Ness KV, Greenstreet TA, March CJ, Kronheim SR, Druck T, Cannizzaro LA, Huebner K, Black RA (1992) Science 256:97
15. Thornberry NA, Bull HG, Calaycay JR, Chapman KT, Howard AD, Kostura MJ, Miller DK, Molineaux SM, Weidner JR, Aunins J, Elliston KO, Ayala JM, Casano FJ, Chin J, Ding GJ-F, Egger LA, Gaffney EP, Limjuco G, Palyha OC, Raju SM, Rolando AM, Salley JP, Yamin T-T, Lee TD, Shively JE, MacCross M, Mumford RA, Schmidt JA, Tocci MJ (1992) Nature 356:768
16. Kumar S, Tomooka Y, Noda M (1992) Biochem Biophys Res Commun 185:1155
17. Kumar S, Kinoshita M, Noda M, Copeland NG, Jenkins NA (1994) Genes Dev 8:1613
18. Wang L, Miura M, Bergeron L, Zhu H, Yuan J (1994) Cell 78:739
19. Xue D, Horvitz HR (1995) Nature 377:248
20. Talanian RV, Quinlan C, Trautz S, Hackett MC, Mankovich JA, Banach D, Ghayur T, Brady KD, Wong WW (1997) J Biol Chem 272:9677
21. Smyth MJ, Trapani JA (1995) Immunol Today 16:202
22. Miura M, Zhu H, Rotello R, Hartwieg EA, Yuan J (1993) Cell 75:653
23. Black RA, Kronheim SR, Sleath PR (1989) FEBS Lett 247:386
24. Kostura MJ, Tocci MJ, Limjuco G, Chin J, Cameron P, Hillman AG, Chartrain NA, Schmidt JA (1989) Proc Natl Acad Sci USA 86:5227
25. Gu Y, Wu J, Faucheau C, Lalanne J-L, Diu A, Livingston DJ, Su MS-S (1995) EMBO J 14:1923
26. Ramage P, Cheneval D, Chvei M, Graff P, Hemmig R, Heng R, Kocher HP, Mackenzie A, Memmert K, Revesz L, Wishart W (1995) J Biol Chem 270:9378
27. Sleath PR, Hendrickson RC, Kronheim SR, March CJ, Black RA (1990) J Biol Chem 265:14526
28. Howard AD, Kostura MJ, Thornberry NA, Ding GJF, Limjuco G, Weidner J, Salley JP, Hogquist KA, Chaplin DD, Mumford RA, Schmidt JA, Tocci MJ (1991) J Immunol 147:2964
29. Rano TA, Timkey T, Peterson EP, Rotonda J, Nicholson DW, Becker JW, Chapman KT, Thornberry NA (1997) Chemistry & Biology 4:149
30. Gu Y, Kuida K, Tsutsui H, Ku G, Hsiao K, Fleming MA, Hayashi N, Higashino K, Okamura H, Nakanishi K, Kurimoto M, Tanimoto T, Flavell RA, Sato V, Harding MW, Livingston DJ, Su MS-S (1997) Science 275:206
31. Ghayur T, Banerjee S, Hugunin M, Butler D, Herzog L, Carter A, Quintal L, Sekut L, Talanian R, Paskind M, Wong W, Kamen R, Tracey D, Allen H (1997) Nature 386:619

32. Wilson KP, Black J-AF, Thomson JA, Kim EE, Griffith JP, Navia MA, Murcko MA, Chambers SP, Aldape RA, Raybuck SA, Livingston DJ (1994) Nature 370:270
33. Walker NPC, Talanian RV, Brady KD, Dang LC, Bump NJ, Ferenz CR, Franklin S, Ghayur T, Hackett CR, Hammill LD, Herzog L, Hugunin M, Houy W, Mankovich J A, McGuiness L, Orlewicz E, Paskind M, Pratt CA, Reis P, Summani A, Terranova M, Welch JP, Xiong L, Moller A, Tracey DE, Kamen R, Wong WW (1994) Cell 78:343
34. Van Criekinge W, Beyaert R, Van de Craen M, Vandenabeele P, Schotte P, De Valck D, Fiers W (1996) J Biol Chem 271:27,245
35. Ray CA, Black RA, Kronheim SR, Greenstreet TA, Sleath PR, Salvesen GS, Pickup DJ (1992) Cell 69:597
36. Gagliardini V, Fernandez P-A, Lee RKK, Drexler HCA, Rotello RJ, Fishman MC, Yuan J (1994) Science 263:826
37. Tewari M, Dixit VM (1995) J Biol Chem 270:3255
38. Enari M, Hug H, Nagata S (1995) Nature 375:78
39. Los M, Van de Craen M, Penning LC, Schenk H, Westendorp M, Baeuerle PA, Droge W, Krammer PH, Fiers W, Schulze-Osthoff K (1995) Nature 375:81
40. Strasser A, Harris AW, Huang DCS, Krammer PH, Cory S (1995) EMBO J 14:6136
41. Miura M, Friedlander RM, Yuan J (1995) Proc Natl Acad Sci USA 92:8318
42. Dorstyn L, Kumar S (1997) Cell Death & Differ (in press)
43. Lazebnik YA, Kaufmann SH, Desnoyers S, Poirier GG, Earnshaw WC (1994) Nature 371:346
44. Chow SC, Weis M, Kass GEN, Holmstrom TH, Eriksson JE, Orrenius S (1995) FEBS Lett 364:134
45. Fearnhead HO, Dinsdale D, Cohen GM (1995) FEBS Lett 375:283
46. Zhivotovsky B, Gahm A, Ankarcrona M, Nicotera P, Orrenius S (1995) Exp Cell Res 221:404
47. Zhou Q, Snipas S, Orth K, Muzio M, Dixit VM, Salvesen GS (1997) J Biol Chem 272:7797
48. Boudreau N, Sympson CJ, Werb Z, Bissell M (1995) Science 267:891
49. Kuida K, Lippke JA, Ku G, Harding MW, Livingston DJ, Su MS-S, Flavell RA (1995) Science 267:2000
50. Li P, Allen H, Banerjee S, Franklin S, Herzog L, Johnston C, McDowell J, Paskind M, Rodman L, Salfeld J, Towne E, Tracey D, Wardwell S, Wei F-Y, Wong W, Kamen R, Seshadri T (1995) Cell 80:401
51. Nagata S (1997) Cell 88:355
52. Enari M, Talanian RV, Wong WW, Nagata S (1996) Nature 380:723
53. Ogasawara J, Watanabe-Fukunaga R, Adachi M, Matzuzawa A, Kasugai T, Kitamura Y, Itoh N, Suda T, Nagata S (1993) Nature 364:806
54. Rodriguez I, Matsuura K, Ody C, Nagata S, Vassali P (1996) J Exp Med 184:2067
55. Faucheu C, Diu A, Chan AWE, Blanchet A-M, Miossec C, Hervé F, Collard-Dutilleul V, Gu Y, Aldape RA, Lippke JA, Rocher C, Su MS-S, Livingston DJ, Hercend T, Lalanne J-L (1995) EMBO J 14:1914
56 Kamens J, Paskind M, Hugunin M, Talanian RV, Allen H, Banach D, Bump N, Hackett M, Johnston CG, Li P, Mankovich JA, Terranova M, Ghayur T (1995) J Biol Chem 270:15,250
57. Munday NA, Vaillancourt JP, Ali A, Casano FA, Miller DK, Molineaux SM, Yamin T-T, Yu VU, Nicholson DW (1995) J Biol Chem 270:15,870
58. Faucheu C, Blanchet A-M, Collard-Dutilleul V, Lalanne JL, Diu-Hercend A (1996) Eur J Biochem 236:207
59. Wang S, Miura M, Jung Y-K, Zhu H, Gagliardini V, Shi L, Greenberg AH, Yuan J (1996) J Biol Chem 271:20,580
60. Fernandes-Alnemri T, Litwack G, Alnemri ES (1994) J Biol Chem 269:30,761
61. Tewari M, Quan LT, O'Rourke K, Desnoyers S, Zeng Z, Beidler DR, Poirier GG, Salvesen GS, Dixit VM (1995) Cell 81:1
62. Nicholson DW, Ali A, Thornberry NA, Vaillancourt JP, Ding CK, Gallant M, Gareau Y, Griffin PR, Labelle M, Lazebnik YA, Munday NA, Raju SA, Smulson ME, Yamin T-T, Yu VL, Miller DK (1995) Nature 376:37
63. Kaufmann SH (1989) Cancer Res 49:5870

64. Rotonda J, Nicholson DW, Fazil KM, Gallant M, Gareau Y, Labelle M, Peterson EP, Rasper DM, Ruel R, Vaillancourt JP, Thronberry NA, Becker JW (1996) Nature Struct Biol 3:619
65. Mittl PRE, Di Marco S, Krebs JF, Bai X, Karanewsky DS, Priestle JP, Tomaselli KJ, Grutter MG (1997) J Biol Chem 272:6539
66. Kuida K, Zheng TS, Na S-Q, Kuan C-Y, Yang D, Karasuyama H, Rakic P, Flavell (1996) Nature 384:368
67. Fernandes-Alnemri T, Litwack G, Alnemri ES (1995) Cancer Res 55:2737
68. Fernandes-Alnemri T, Takahashi A, Armstrong R, Krebs J, Fritz L, Tomaselli KJ, Wang L, Yu Z, Croce C, Salvesen G, Earnshaw WC, Litwack G, Alnemri ES (1995) Cancer Res 55:6045
69. Duan H, Chinnaiyan AM, Hudson PL, Wing JP, He W-W, Dixit VM (1996) J Biol Chem 271:1621
70. Lippke JA, Gu Y, Sarnecki C, Caron PR, Su MS-S (1996) J Biol Chem 271:1825
71. Lazebnik YA, Takahashi A, Moir RD, Goldman RD, Poirier GG, Kaufmann SH, Earnshaw WC (1995) Proc Natl Acad Sci USA 92:9042
72. Orth K, Chinnaiyan AM, Garg M, Froelich CJ, Dixit VM (1996) J Biol Chem 271:16,443
73. Schlegel J, Peters I, Orrenius S, Miller DK, Thornberry NA, Yamin TT, Nicholson DW (1996) J Biol Chem 271:1841
74. Chinnaiyan AM, Orth K, O'Rourke K, Duan H, Poirier GG, Dixit VM (1996) J Biol Chem 271:4573
75. Mukasa T, Urase K, Momoi MY, Kimura I, Momoi T (1997) Biochem Biophys Res Comm 231:770
76. Martins LM, Kottke T, Mesner PW, Basi GS, Sinha S, Frigon N Jr, Tatar E, Tung JS, Bryant K, Takahashi A, Svingen PA, Madden BJ, McCormick DJ, Earnshaw WC, Kaufmann SH (1997) J Biol Chem 272:7421
77. Keane RW, Srinivasan A, Foster LM, Testa MP, Ord T, Nonner D, Wang HG, Reed JC, Bredesen DE, Kayalar C (1997) J Neurosci Res 48:168
78. MacFarlane M, Cain K, Sun X-M, Alnemri ES, Cohen GM (1997) J Cell Biol 137:469
79. Faleiro L, Kobayashi R, Fearnhead H, Lazebnik Y (1997) EMBO J 16:2271
80. Harvey NL, Butt AJ, Kumar S (1997) J Biol Chem 272:13,134
81. Fernandes-Alnemri T, Armstrong RC, Kreb J, Srinivasula SM, Wang L, Bullrich F, Fritz LC, Trapani JA, Tomaselli KJ, Litwack G, Alnemri ES (1996) Proc Natl Acad Sci USA 93:7464
82. Boldin MP, Goncharov TM, Goltsev YV, Wallach, D (1996) Cell 85:803
83. Muzio M, Chinnaiyan AM, Kischkel FC, O'Rourke K, Shevchenko A, Scaffidi C, Bretz JD, Zhang M, Ni J, Gentz R, Mann N, Krammer PH, Peter ME, Dixit VM (1996) Cell 85:817
84. Duan H, Orth K, Chinnaiyan AM, Poirier GG, Froelich CJ, He W-W, Dixit VM (1996) J Biol Chem 271:16,720
85. Srinivasula SM, Fernandes-Alnemri T, Zangrilli J, Robertson N, Armstrong RC, Wang L, Trapani JA, Tomaselli KJ, Litwack G, Alnemri ES (1996) J Biol Chem 271:27,099
86. Hsu H, Xiong J, Goeddel DV (1995) Cell 81:495
87. Vincenz C, Dixit VM (1997) J Biol Chem 272:6578
88. Muzio M, Salvesen GS, Dixit VM (1997) J Biol Chem 272:2952
89. Allet B, Hochmann A, Martinou I, Missotten M, Antonsson B, Sadoul R, Martinou J-C, Bernasconi L (1996) J Cell Biol 135:480
90. Kumar S (1995) FEBS Lett 368:69
91. Troy CM, Stefanis L, Greene LA, Shelanski ML (1997) J Neurosci 17:1911
92. Kumar S, Kinoshita M, Dorstyn L, Noda M (1997) Cell Death & Differ (in press)
93. Kinoshita M, Tomimoto H, Kinoshita A, Kumar S, Noda, M (1997) J Cereb Blood Flow Metab (in press)
94. Asahi M, Hoshimaru M, Uemura Y, Tokime T, Kojima M, Ohtsuka T, Matsuura N, Aoki T, Shibahara K, Kikuchi H (1997) J Cereb Blood Flow Metab 17:11
95. Flaws JA, Kuju K, Trbovich AM, DeSanti A, Tilly KI, Hirshfield AN, Tilly JL (1995) Endocrinology 136:5042
96. Duan H, Dixit VM (1997) Nature 385:86
97. Ahmad M, Srinivasula SM, Wang L, Talanian RV, Litwack G, Fernendes-Alnemri T, Alnemri ES (1997) Cancer Res 57:615

98. Chinnaiyan AM, O'Rourke K, Yu G-L, Lyons RH, Garg M, Duan DR, Xing L, Gentz R, Ni J, Dixit VM (1996) Science 274:990
99. Kiston J, Raven T, Jiang Y-P, Goeddel DV, Giles KM, Pun K-T, Grinham CJ, Brown R, Farrow SN (1996) Nature 384:372
100. Wang Z-Q, Aeur B, Sting L, Berghammer H, Haidacher D, Schweiger M, Wagner EF (1995) Genes Dev 9:509
101. Casciola-Rosen LA, Miller DK, Anhalt GJ, Rosen A (1994) J Biol Chem 269:30,757
102. Tewari M, Beidler DR, Dixit VM (1995) J Biol Chem 270:18,738
103. Casciola-Rosen LA, Anhalt GJ, Rosen A (1995) J Exp Med 182:1625
104. Song Q, Lees-Miller SP, Kumar S, Zhang N, Chan DW, Smith GCM, Jackson SP, Alnemri ES, Litwack G, Khanna KK, Lavin MF (1996) EMBO J 15:3238
105. Song Q, Burrows S, Lees-Miller SP, Smith G, Jackson S, Kumar S, Trapani JA, Alnemri ES, Litwack G, Lu H, Moss D, Lavin MF (1996) J Exp Med 184:619
106. McConnell KR, Dynan WS, Hardin JA (1997) J Immunol 158:2083
107. Wang X, Pai J-T, Wiedenfeld EA, Medina JC, Slaughter CA, Goldstein JL, Brown MS (1995) J Biol Chem 270:18,044
108. Wang X, Zelenski NG, Yang J, Sakai J, Brown MS, Goldstein JL (1996) EMBO J 15:1012
109. Pai J-T, Brown MS, Goldstein JL (1996) Proc Natl Acad Sci USA 93:5437
110. Ucker DS, Obermiller PS, Eckhardt W, Apgar JR, Berger NA, Meyers J (1992) Mol Cell Biol 12:3060
111. Oberhammer FA, Hochegger K, Froschl G, Tiefenbacher R, Pavelka M (1994) J Cell Biol 126:827
112. Kayalar C, Ord T, Testa MP, Zhong L-T, Bredesen DE (1996) Proc Natl Acad Sci USA 93:2234
113. Mashima T, Naito M, Noguchi K, Miller DK, Nicholson DW, Tsuruo T (1996) Oncogene 14:1007
114. Martin SJ, O'Brien GA, Nishioka WK, McGahon AJ, Mahboubi A, Saido TC, Green DR (1995) J Biol Chem 270:6425
115. Vanags DM, Porn-Ares I, Coppola S, Burgess DH, Orrenius S (1996) J Biol Chem 271:31,075
116. Cryns VL, Bergeron L, Zhu H, Li H, Yuan J (1996) J Biol Chem 271:31,277
117. Brancolini C, Benedetti M, Schneider C (1995) EMBO J 14:5179
118. Emoto Y, Manome Y, Meinhardt G, Kisaki H, Kharbanda S, Robertson M, Ghayur T, Wong WW, Kamen R, Weichselbaum R, Kufe D (1995) EMBO J 14:6148
119. Na S, Chuang T-H, Cunningham A, Turi TG, Hanke JH, Bokoch GM, Danley DE (1996) J Biol Chem 271:11,209
120. Beyaert R, Kidd VJ, Cornelis S, Van de Craen M, Denecker G, Lahti JM, Gururajan R, Vandenabeele P, Fiers W (1997) J Biol Chem 272:11,694
121. Janicke RU, Walker PA, Lin XY, Porter AG (1996) EMBO J 15:6969
122. Chen WD, Otterson GA, Lipkowitz S, Khleif SN, Coxon AB, Kaye FJ (1997) Oncogene 14:1243
123. Tan X, Martin SJ, Green DR, Wang JYJ (1997) J Biol Chem 272:9613
124. Waterhouse N, Kumar S, Song Q, Strike P, Sparrow L, Dreyfuss G, Alnemri ES, Litwack G, Lavin MF, Watters D (1996) J Biol Chem 271:29,335
125. Song Q, Lu H, Zhang N, Luckow B, Shah G, Poirier G, Lavin M (1997) Biochem Biophys Res Commun 233:343
126. Wissing D, Mouritzen H, Egeblad M, Poirier GG, Jaattela M (1997) Proc Natl Acad Sci USA 94:5073
127. Liu X, Zou H, Slaughter C, Wang X (1997) Cell 89:175
128. Berke G (1995) Cell 89:9
129. Vaux DL, Haecker G, Strasser A (1994) Cell 78:777
130. Harvey NL, Trapani JA, Fernandes-Alnemri T, Litwack G, Alnemri ES, Kumar S (1996) Genes Cells 1:673
131. Darmon AJ, Nicholson DW, Bleackley RC (1995) Nature 377:446
132. Darmon AJ, Ehrman N, Caputo A, Fujinaga J, Bleackley RC (1994) J Biol Chem 269:32,043

133. Anel A, Gamen S, Alava MA, Schmitt-Verhulst AM, Pineiro A, Naval J (1997) J Immunol 158:1999
134. Bump NJ, Hackett M, Hugunin M, Seshagiri S, Brady K, Chen P, Ferenz C, Franklin S, Ghayur T, Li P, Licari P, Mankovich J, Shi L, Greenberg AH, Miller LK, Wong WW (1995) Science 269:1885
135. Clem RJ, Fechheimer M, Miller LJ (1991) Science 254:1388
136. Sugimoto A, Friesen PD, Rothman JH (1994) EMBO J 13:2023
137. Rabizadeh S, LaCount DJ, Friesen PD, Bredesen DE (1993) J Neurochem 61:2318
138. Martinou I, Fernandez P-A, Missotten M, White E, Allet B, Sadoul R, Martinou J-C (1995) J Cell Biol 128:201
139. Beidler DR, Tewari M, Friesen PD, Poirier G, Dixit VM (1995) J Biol Chem 270:16,526
140. Bertin T, Mendrysa SM, Lacount DJ, Gaur S, Krebs JF, Armstrong RC, Tomaselli KJ, Friesen PD (1996) J Virol 70:6251
141. Bertin J, Armstrong RC, Ottilie S, Martin DA, Wang Y, Banks S, Wang G-H, Senkevich TG, Alnemri ES, Moss B, Lenardo MJ, Tomaselli KJ, Cohen JI (1997) Proc Natl Acad Sci USA 94:1172
142. Thome M, Schneider P, Hofmann K, Fickenscher H, Meinl E, Neipel F, Mattmann C, Burns K, Bodmer J-L, Schroter M, Scaffidi C, Krammer PH, Peter ME, Tschopp J (1997) Nature 386:517
143. Chinnaiyan AM, Orth K, O'Rourke K, Duan H, Poirier GG, Dixit VM (1996) J Biol Chem 271:4573
144. Erhardt P, Cooper GM (1996) J Biol Chem 271:17,601
145. Srinivasan A, Foster LM, Testa MP, Ord T, Keane RW, Bredesen DE, Kayalar C (1996) J Neurosci 16:5654
146. Perry DK, Smyth MJ, Wang H-G, Reed JC, Poirier GG, Obeid LM, Hannun YA (1997) Cell Death & Differ 4:29
147. Estoppey S, Rodriguez I, Sadoul R, Martinou J-C (1997) Cell Death & Differ 4:34
148. White E (1996) Genes Dev 10:1
149. Oltvai ZN, Millman CL, Korsmeyer SJ (1993) Cell 74:609
150. Oltvai ZN, Korsmeyer SJ (1994) Cell 79:189
151. Spector MS, Desnoyers S, Hoeppner DJ, Hengartner MO (1997) Nature 385:653
152. Chinnaiyan AM, O'Rourke K, Lane BR, Dixit VM (1997) Science 275:1122
153. Wu D, Herschel HD, Nunez G (1997) Science 275:1126
154. Liu X, Kim CN, Yang J, Jemmerson R, Wang X (1996) Cell 86:147
155. Yang J, Liu X, Bhalla K, Kim CN, Ibrado AM, Cai J, Peng T-I, Jones DP, Wang X (1997) Science 275:1129
156. Kluck RM, Bossy-Wetzel E, Green DR, Newmeyer DD (1997) Science 275:1132
157. Zou H, Henzel WJ, Liu X, Lutschg A, Wang X (1997) Cell 90:405

Received November 1997

"Tissue" Transglutaminase and Apoptosis

Francesco Autuori · Maria Grazia Farrace · Serafina Oliverio · Lucia Piredda · Mauro Piacentini

Department of Biology, University of Rome "Tor Vergata", Via della Ricerca Scientifica 00133 Rome Italy

In this paper we discuss the role of "tissue" transglutaminase (tTG) in apoptosis. This enzyme by catalizing the Ca^{2+}-dependent cross-linking of intracellular proteins leads to the formation of the SDS-insoluble protein scaffold in cells undergoing programmed cell death. These intracellular structures confer resistance to mechanical and chemical attack to the polipeptides involved in the linkages. tTG is induced during apoptosis, in fact, tTG mRNA is trascripted as a consequence of apoptosis induction. Overexpression of tTG in many cell lines enhances their susceptibility to apoptosis, indicating a pivotal role for tTG in this process. In keeping with these findings transfection of the human tTG complementary DNA in antisense orientation leads in a pronounced decrease of both spontaneous as well as induced apoptosis. Interestingly, the identification of the tTG substrate proteins in cells undergoing apoptosis has evidenced that many of the tTG proteins are also substrates of caspases.

Keywords: Protein cross-links, Cell Death, Retinoblastoma protein, Caspases, Calpains.

List of Symbols and Abbreviations

DAG	1,2-diacylglycerol
ICE	interleukin 1β converting enzyme
IL6	interleukin 6
IP_3	inositol-1,4,5-triphosphate
PGE_2	prostaglandin E2
PKC	Protein kinase C
PLC	phospholipase C

Advances in Biochemical Engineering / Biotechnology, Vol. 62
Managing Editor: Th. Scheper
© Springer-Verlag Berlin Heidelberg 1998

TCR   T cell receptor
TGFβ tumor growth factor
tTG   "tissue" transglutaminase

1
Introduction

The transglutaminases family comprises intracellular and extracellular enzymes catalyzing Ca^{2+}-dependent reactions resulting in the post-translational modification of proteins at the level of glutamine and lysine residues [1–3]. This post-translational modification might lead to the formation of the $\varepsilon(\gamma$-glutamyl)-lysine cross-linkings and/or to the covalent incorporation of di- and polyamines and histamine [1–3]. Diamines and polyamines may also participate in cross-linking reactions through the formation of N,N-bis(γ-glutamyl)polyamine bonds [3, 4]. The formation of these covalent cross-links leads to oligomerization of substrate protein which acquires peculiar features of resistance to breakage and chemical attack [1]. The polypeptides can be released from the polymer(s) only by the proteolytic degradation of protein chains [1, 5, 6]. In fact, endoproteases capable of hydrolyzing the cross-links formed by transglutaminases have not been described in vertebrates [1, 7, 8]. In keeping with this, the assembly of cross-linked protein polymers does not take place in proliferating cells, but occurs in terminally differentiated cells such as keratinocytes and chondrocytes as well as in cells undergoing death by apoptosis [6, 8–10]. The various transglutaminase forms seem to be involved in different biological phenomena (blood coagulation, wound healing, keratinocyte terminal differentiation and cell death by apoptosis) which are related to the protection of cell and tissue homeostasis [1, 8, 10].

2
"Tissue" Transglutaminase as a G-Protein

"Tissue" transglutaminase (tTG) or type-II transglutaminase gene encodes for a protein which, in mammals, has a molecular weight of about 80 kDa [8, 10]. TG gene is constitutively expressed in a few cell types localized in peculiar sites of mammalian tissues (endothelial cells, smooth muscle cells and mesangial cells) [11]. tTG catalytic activity is potentiated by calcium ions and is specifically inhibited by guanine nucleotides (GTP) [12]. In keeping with this, recent studies have identified a new potential biological role of tTG [12]; in fact, it has been demonstrated that tTG is the $G_{\alpha h}$ subunit associated with the 50 kDa β subunit ($G_{\beta h}$) of the GTP-binding protein (G_h). The G_h dimer acts in association with the rat liver α_1-adrenergic receptor in a ternary complex. Thus, the $G_{\alpha h}$ is a multifunctional protein which, by binding GTP in a $G_{\alpha h GTP}$ complex, can modulate receptor-stimulated phospholipase-C (PLC) activation. $G_{\alpha h}$ represents a novel class of GTP-binding proteins that participate in the receptor-mediated signaling pathway. It is important to note that the GTP binding activity of tTG/$G_{\alpha h}$ actively prevents activation of the cross-linking activity of tTG. In fact, the Ca^{2+}-

dependent cross-linking activity of tTG is finely tuned by GTP binding levels which in turn regulate secondary messengers such as the production of inositol-1,4,5-triphosphate (IP$_3$) and *sn*-1,2-diacylglycerol (DAG) from phosphatidyl-inositol-4,5-biphosphate (PIP$_2$).

3
"Tissue" Transglutaminase and Apoptosis

3.1
tTG Expression and Regulation in Dying Cells

Under physiological conditions tTG gene is not expressed in the majority of cells and its mRNA is transcribed as a consequence of the induction of apoptosis [13]. The onset of apoptosis is generally associated with a large increase of the tTG-mRNA level followed by an enhancement of the enzyme synthesis and of the cross-linking activity [13–17]. Although tTG is not active at the Ca^{2+} levels present in viable cells, the increase of the intracellular Ca^{2+} concentration reported in cells undergoing apoptosis is sufficient to activate the enzyme [1, 7, 13–16]. The activation of tTG protein in the dying cells results in the assembly of a highly cross-linked intracellular protein net stabilized by both spermidine-derived and $\varepsilon(\gamma$-glutamyl)lysine cross-links [7, 15]. The experimental models in which the enzyme has been shown to increase in apoptotic cells include the best characterized *in vivo* and *in vitro* cell death systems (for review see [10, 18]).

It is not known what role is played by tTG in those cells which constitutively express the enzyme. However, the fine modulation of the tTG/G$_{\alpha h}$ protein by GTP and Ca^{2+} (and possibly additional molecules such as free putrescine and other polyamines) may explain how cells are able to survive in the presence of high tTG/G$_{\alpha hGTP}$ protein levels in their cytoplasm. Cell death prevention could also be achieved by the DAG-dependent activation of protein kinase C (PKC; [19]. Nevertheless, cells constitutively expressing tTG are localized in tissue areas exposed to environmental and functional stress [11]; hence, to avoid harmful consequences, they might have the apoptotic machinery in place, ready to act whenever their integrity is affected. In fact, in some *in vitro* settings, apoptosis can take place in the absence of protein synthesis in anucleated cells [20]. Thus, it is tempting to speculate that in the course of evolution the tTG gene has acquired a cell-type-specific regulation, which allows the preventive accumulation of the enzyme in cells particularly exposed to environmental stress. It must be recalled that the tTG protein can be post-translationally regulated and is inactive in microenvironments which have low Ca^{2+} and high GTP levels [21]. These findings raise the important question of the regulation of the effector genes during the execution phase of apoptosis. It is well known that apoptosis can be induced in different cells by a wide range of physiological as well as pathological stimuli [8, 22]. This finding could imply that regulation of the putative "killer" genes has multiple accesses or that different lethal stimuli end in a common signalling pattern. It has been shown that *de novo* transcription of the tTG gene is induced by several factors (retinoic acid, PGE2, IL6, TGFβ) which also modulate apoptosis [8, 10, 21, 23]. This multiple regulation might be typical of the effector elements of the cell death program. The recent cloning and sequencing of the 5′ flanking region of the tTG gene seems to confirm this hypothesis. In fact, the sequence analysis of fragments of the cloned genomic DNA revealed the presence of potential binding sites for several regulatory factors [23, 24], thus suggesting that the transcription of a putative apoptotic "effector" gene can be controlled by a multifunctional promoter.

3.2
tTG Protein Substrates in Apoptosis

Cells undergoing apoptosis are characterized by typical ultrastructural modifications: cell shrinkage, nuclear collapse associated to chromatin degradation and condensation, alterations of the cytoskeleton and phagocytosis of the resulting apoptotic bodies [22, 25]. These changes occur in both unicellular organisms and mammalian cells regardless of the stimulus that leads the cell into apoptosis [25], suggesting the presence of cellular targets which are invariably modified to establish the death phenotype. The identification of one of the killer genes of the *C. elegans* genetic death pathway, ced-3 [26], as the cysteine protease interleukin-1β converting enzyme (ICE) has led to the discovery of many other cysteine proteases (recently classified as caspases) as well as to the calpains involvement in apoptosis [25, 27]. A large number of protein substrates have been shown to be cleaved by these thiol proteases during apoptosis, although, it is still unclear how many of these protein substrates must be processed to establish the death phenotype [25, 27].

The biochemical characterization of the cross-linked protein polymers formed by tTG in dying cells has revealed the presence of several intracellular proteins (actin, annexin, vinculin, fibronectin, involucrin and some unknown proteins [28–30]). Recently, the retinoblastoma protein (pRB) and troponin have also been shown to undergo tTG-dependent polymerization in apoptosis [31–33] (Table 1). Most of these proteins are cleaved by caspases and calpains during the cell death execution phase [27, 30, 32, 33], thus indicating that tTG and thiol proteases process the same set of target proteins. In keeping with this hypothesis, it has been shown that the histones which are cleaved during apoptosis [27] act as tTG substrates [34] (Table 1). The finding that the putative killer proteins involved in apoptosis act on actin, troponin, vinculin, histone H2B and pRB identifies these proteins as the primary "death targets" (Table 1). Although the physiological significance of the "cleaving and polymerizing" of the same substrate proteins in the establishment of the apoptotic phenotype has yet to be defined, a few interesting observations can be made. The presence of a cysteine active-site is essential for the catalytic activity of tTG, calpains and caspases [27, 35]. tTG has recently been shown not only to promote the formation of protein cross-links, but also to cleave them, acting as a hydrolytic enzyme [35]. This finding supports the biochemical similarities existing between tTG and papain, a thiol protease from which tTG is likely derived [35]. Considering that papain, like caspases, cleaves proteins at the *asp* residues [27, 35], these observations suggest that a papain-like ancestral thiol protease might have been selected as the original cell "executioner". This gene might have evolved into the

Table 1. Proteins post-translationally modified by both tTG and thiol proteases during apoptosis

tTG substrate	thiol protease substrate
histone H2B [34]	caspase [62]
pRB [31]	caspase [32]
actin [30]	caspase [63]
tubulin [61]	calpain [25]
troponin [33]	calpain [33]
vinculin [60]	calpain [60]

different classes of sophisticated executioners (caspases, calpains and tTG) able to drive the highly regulated form(s) of death currently being studied in multicellular organisms.

3.3
Is tTG an Essential Effector Element of Apoptosis?

A definitive role for tTG in apoptosis has not yet been firmly established, although tTG expression does not seem to be a late epiphenomenon. Interesting clues, supporting this assumption, derive from transfection studies carried out in various mammalian cells. Human neuroblastoma SK-N-BE(2), BALB-C 3T3 and L929 fibroblasts transfected with a full length tTG cDNA show a large reduction in their proliferative capacity paralleled by an increased spontaneous cell death rate [36–38]. The dying tTG-transfected cells exhibit both cytoplasmic and nuclear changes characteristic of cells undergoing apoptosis. Conversely, transfection of neuroblastoma cells with an expression vector containing segments of the human tTG complementary DNA in antisense orientation results in a pronounced decrease of both spontaneous as well as induced apoptosis [37]. These findings indicate in the tTG-catalyzed irreversible cross-linking of intracellular protein an important biochemical event in the induction of the structural changes featuring cells dying by apoptosis. The question arises as to whether in the early stages of death pathway a regulated tTG-mediated post-translational modification of specific protein substrate(s) might have a role in the commitment to apoptosis. We have recently shown that in U937 cells the polymerization of pRB, which precedes apoptosis, is mediated by its post-translational modification catalyzed by tTG [31]. Thus, suggesting that the activation of this cross-linking enzyme might determine an irreversible commitment to death. pRB plays a key role in cell cycle control [39, 40]. However, it has also been shown that the absence of functional pRB may result in apoptosis rather than in uncontrolled cell proliferation [41–43] and that overexpression of functional pRB may induce apoptosis or rescue cells from death, depending on the system [44–46]. Homozygous pRB-null mice die during gestation showing massive induction of apoptosis during liver erythropoiesis and neuronal development [40]. The question arises as to the functional implications of the tTG-dependent polymerization of pRB occurring during the early phases of apoptosis. In keeping with the possibility that pRB polymerization might lead to its functional inactivation, we showed that the tTG-dependent polymerization of pRB is paralleled by the rapid disappearance of E2F-1, which occurs when the transcription factor is not protected by pRB binding from entering the Ub/proteasome pathway [47, 48].

4
Concluding Remarks

The original idea that cell death by apoptosis was a genetically-regulated event has been confirmed by the identification of several genes participating in the process [7, 49–59]. However, in spite of the exponential increase in the number of studies on gene-dependent cell death, a single "killer" gene has not yet been

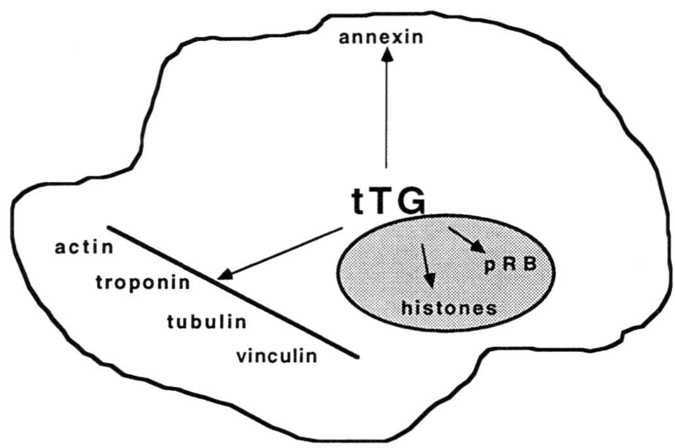

Fig. 1.

identified in mammalian cells. A number of distinct enzymes (i.e. caspase, calpains and tTG) might work in a coordinate fashion to achieve the irreversible, fast and clean removal of apoptotic bodies. Under controlled physiological conditions it is very likely that different effector elements play complementary integrated functions in different cell compartments.

We propose the tTG gene product as a candidate "killer" element of the apoptotic program. In fact, the findings discussed in this review indicate that tTG might have more than one function within the cascade of events leading to the establishment of the apoptotic phenotype: i) an early regulatory function which, through the polymerization of pRB, may influence the decision to undergo apoptosis or survive (Fig. 1); ii) a late effect when tTG in cooperation with other effector elements might have a direct effect in killing and/or tTG-dependent cross-linking could stabilize the apoptotic cells before their clearance (Fig. 1).

Acknowledgements. This work was supported by grants from E.C. Biotechnology "IV framework", A.I.R.C. and C.N.R. Dr. Oliverio was supported by an A.I.R.C. grant.

5
References

1. Folk JE (1980) Transglutaminases. Annu Rev Biochem 49:517
2. Fesus L, Szucs EF, Barrett KE, Metcalfe DD, Folk JE (1985) J Biol Chem 260:13,771
3. Piacentini M, Martinet N, Beninati S, Folk JE (1988) J Biol Chem 263:3790
4. Folk JE, Park MH, Chung SI, Schrode J, Lester EP, Cooper HL (1980) J Biol Chem 255:3695
5. Folk JE, Finlayson S (1977) Adv Prot Chem 31:1
6. Green H (1977) Cell 11:405
7. Fesus L, Thomazy V, Autuori F, Ceru' MP, Tarcsa E, Piacentini M (1989) FEBS Lett 245:150
8. Fesus L, Davies PJA, Piacentini M (1991) Europ J Cell Biol 56:170
9. Aeschlimann D, Wetterwald A, Fleisch H, Paulsson M (1993) J Cell Biol 120:1461–1470
10. Piacentini M, Davies PJA, Fesus L (1994) In: Tomei LD, Cope FO (eds) Apoptosis II: the molecular basis of apoptosis in disease. Cold Spring Harbor Laboratory Press
11. Thomazy V, Fesus L (1989) Cell Tissue Res 255:215
12. Nakaoka, H, Perez DM, Baek KJ, Das T, Husain A, Misono K, Im M, Graham RM (1994) Science 264:1593
13. Piacentini M, Autuori F, Dini L, Farrace MG, Ghibelli L, Piredda L, Fesus L (1991) Cell Tissue Res 263:227
14. Fesus L, Thomazy V, Falus A (1987) FEBS Lett 224:104
15. Piacentini M, Fesus L, Farrace MG, Ghibelli L, Piredda P, Melino G (1991) Eur J Cell Biol 54:246
16. Knight CRL, Rees RC, Griffin M (1991) Biochim Biophys Acta, 1096:312
17. Strange R, Li F, Saurer S, Burkhard A, Friis RR (1992) Development 115:49
18. Piacentini M (1995) Curr Top Microbiol Immunol 200:163
19. Bruns RF, Miller FD, Merruman RL, Howbert JJ, Heath WH, Kobayashi E, Takahashi I, Tamaoki T, Nakano H (1991) Biochem Biophys Res Commun 176:288
20. Raff MC, Barres BA, Burne JF, Coles HS, Ishizaki Y, Jacobson MD (1992) Science 262:695
21. Greenberg CS, Birckbichler P, Rice RH (1992) FASEB J 5:3071
22. Arends MJ, Wyllie AH (1991) Int Rev Exp Pathol 32:223
23. Suto N, Ikura K, Shinagawa R, Sasaki R (1993) Biochim Biophys Acta 1172:319
24. Nagy L, Saydak M, Shipley N, Lu S, Basilion JP, Yua Yan Z, Syka P, Chandraratna RA, Stein JP, Heyman RA, Davies PJA (1996) J Biol Chem 23:4355
25. Hale AJ, Smith CA, Sutherland LC, Stoneman VEA, Longthorne VL, Culhane AC, Williams GT (1996) Europ J Biochem 236:1
26. Yuan J, Shaham S, Ledoux S, Ellis HM, Horvitz HR (1993) Cell 75:641

27. Kumar S, Lavin MF (1996) Cell Death Differ 3:255
28. Knight CRL, Hand D, Piacentini M, Griffin M (1993) Europ J Cell Biol 60:210
29. Tarcsa E, Kedei N, Thomazy V, Fesus L (1993) J Biol Chem 267:25,648
30. Nemes Z, Adany R, Balazs M, Boross P, Fesus L (1997) J Biol Chem 272:20577
31. Oliverio S, Amendola A, Di Sano F, Farrace MG, Fesus L, Nemes Z, Piredda L, Spinedi A, Piacentini M (1997) Mol Cell Biol 17:6040
32. Janicke RU, Walker PA, Lin XY, Porter AG (1996) EMBO J 15:6969
33. Gorza L, Menabò L, Di Lisa F, Vitadello M (1997) Am J Pathol 150:2087
34. Ballesteras E, Abad C, Franco L(1996) J Biol Chem 271:18,817
35. Parameswaran KN, Cheng X-F, Chen EC, Velasco PT, Wilson JH, Lorand L (1997) J Biol Chem 272:10,311
36. Gentile V, Thomazy V, Piacentini M, Fesus L, Davies PJA (1992) J Cell Biol 119:463
37. Melino G, Annicchiarico-Petruzzelli M, Piredda L, Candi E, Gentile V, Davies PJA, Piacentini M (1994) Mol Cell Biol 14:6584
38. Piredda L, Amendola A, Colizzi V, Davies PJA, Farrace MG, Fraziano M, Gentile V, Uray I, Piacentini M, Fesus L (1997) Cell Death Differ 4:463
39. DeCaprio JA, Ludlow JW, Dennis L, Furukawa Y, Griffin J, Piwnica-Worms S, Huang H, Livingstone DM (1989) Cell 58:1085
40. Weinberg RA (1995) Cell 81:323
41. Almasan A, Yin Y, Kelly Lee RE, Bradley A, Bertino JR, Li W, Wahl GM (1995) Proc Natl Acad Sci USA 92:5436
42. Morgenbesser SD, Williams BO, Jacks TK, DePinho RA (1994) Nature 371:72
43. Slack RS, Skerjanc IS, Lach B, Craig J, Jardine K, McBurney MW (1995) J Cell Biol 129:77,941
44. Haas-Kogan DA, Kogan SC, Levi D, Dazin T'Ang PA, Fung Y-KT, Israel MA (1995) EMBO J 14:46,142
45. Martel C, Batschè E, Harper F, Cremisi C (1996) Cell Death Differ 3:28,543
46. Xu H-J, Xu K, Zhou Y, Li J, Benedict WF, Hu S-X (1994) Proc Natl Acad Sci USA 91:9837
47. Hateboer G, Kerkhoven RM, Shvarts A, Bernards R, Beijersbergen RL (1996) Genes & Dev 10:2960
48. Hofmann F, Martelli F, Livingston DM, Wang Z (1996) Genes & Dev 10:2949
49. Buttyan R, Olsson CA, Pintar J, Chang C, Bandyk M, Poying NG, Sawczuk IS (1989) Mol Cell Biol 9:3473
50. Bursch, W, Oberhammer F, Schulte-Hermann R (1992) TIPS 13:245
51. Hockenbery D, Zutter M, Hickey W, Nahm M, Korsmeyer SJ (1990) Nature 348:334
52. Savill JS, Dransfield I, Hogg A, Haslett C (1990) Nature 343:170
53. Ellis RE, Yuan J, Horvitz HR (1991) Annu Rev Cell Biol 7:663
54. Yonish-Rouach ED, Resnitzky S, Lotem J, Sachs S, Kimchi G, Oren M (1991) Nature 352:345
55. Dini L, Autuori F, Lentini A, Oliverio S, Piacentini M (1992) FEBS Lett 296:174
56. Evan GI, Wyllie AH, Gilbert CS, Littlewood TD, Land H, Brooks M, Waters CM, Penn LZ, Hancock DC (1992) Cell 69:119
57. Miura M, Zhu H, Rotello R, Hartwieg EA, Yuan J (1993) Cell 75:653
58. Shi L, Nishioka WK, Th'ng J, Bradbury EM, Litchfield DW, Grenberg AH (1993) Science 263:1143
59. Peitsch MC, Polzar B, Stephan H, Crompton T, MacDonald HR, Mannherz HG, Tschopp J (1993) EMBO J 12:371
60. Winzler C, Rovere P, Rescigno M, Granucci F, Penna G, Adorini L, Zimmermann VS, Davoust J, Riccardi-Castagnoli P (1997)J Exp Med 185:317
61. Ambron RT, Kremzner LT (1982) Proc Natl Acad Sci (USA) 79:3442
62. Baxter GD, Smith DJ, Lavin MF (1989) Biochim Byophys Res Commun 162:30
63. Kayalar C, Ord T, Testa MP, Zhong LT, Bredesen DE (1996) Proc Natl Acad Sci (USA) 93:2234

Received January 1998

Survival Factors and Apoptosis

Rosemary O'Connor

Department of Biochemistry, University College Cork, Lee Maltings, Cork, Ireland,
E-mail: roc@ucc.ie

This chapter will explore the role of survival factors in suppression of apoptosis, and illustrate how survival signals play a critical role in the survival of both normal and tumor cells. Survival factors necessary for the development and maintenance of the nervous system and hemopoietic system will be surveyed. This will be followed by a detailed discussion of the role of insulin-like growth factor I (IGF-I) and its receptor in suppression of apoptosis. The importance of survival signals from the IGF-IR for development and tumorigenesis will be discussed, and results of a mutational analysis of the receptor to assign domains necessary for suppression of apoptosis will be summarized. Finally, a discussion of the signal transduction pathways involved in survival factor-signaling will review the roles played by PI-3 kinase and AKT and speculate on how activation of these kinases by survival factors might regulate the apoptotic pathway.

Keywords: Apoptosis, Survival factors, IGF-I, IGF-I receptor, Signal transduction, AKT, PI-3 kinase.

Advances in Biochemical Engineering/
Biotechnology, Vol. 62
Managing Editor: Th. Scheper
© Springer-Verlag Berlin Heidelberg 1998

List of Symbols and Abbreviations

BFU-E   burst forming unit-erythroid
CFU-E   colony forming unit erythroid
GM-CSF   granulocyte macrophage colony stimulating factor
IL-2   interleukin-2
IL-3   interleukin-3

1
Suppression of Apoptosis is Required for Cell Survival

1.1
Apoptosis

Apoptosis or cell death is a genetically controlled and highly regulated process
by which cells die in a variety of physiological and stress-induced situations. It
is characterized by recognizable morphological changes in the cell such as
shrinkage, blebbing, and condensation of the nucleus with internucleosomal
fragmentation of the DNA [1, 2]. Accumulating evidence indicates that the execution of apoptosis is mediated by a family of cysteine proteases or caspases,
(reviewed in [3]) which includes the C.elegans protein CED-3 [4], interleukin 1-β
converting enzyme (ICE or caspase 1) [5] and CPP32 or caspase 3 [6]. A number
of regulators of apoptosis have also been identified in most cell systems. These
include the Fas/TNF/TRAIL receptor family [7–12] which are activated by a

range of soluble and cell bound ligands; the Bcl-2 family of proteins [13] such as the anti-apoptotic proteins Bcl-2 [14] and Bcl-xl [15] and the pro-apoptotic proteins Bax [16] and Bak [17]; and survival factors such as IGF-I [18].

The more we learn about the regulation and execution of the cell death program the more apparent it becomes that this program is a default pathway that can be accessed during the critical decisions that cells make, such as during development, during cell division, in response to viral infection or oncogene activation, and in response to toxic stimuli. All of the effector proteins and regulator proteins necessary for cells to undergo apoptosis are already present in cells, but are kept in check by anti-apoptotic signals such as those delivered by survival factors and anti-death genes.

1.2
Suppression of Apoptosis in *C. elegans*

The necessity for the maintenance of cell survival by suppression of apoptosis has been conserved during evolution, and is well illustrated in the nematode *C. elegans*, where the genetic pathway governing cell death has been extensively studied [19]. Fourteen genes associated with cell death in the worm have been identified (reviewed in [20]) The regulators and effectors of cell death in the worm are families of proteins which are homologous to the Bcl-2 family and caspase family members [4, 21–23]. In this system 131 cells out of a total of 1090 somatic cells are programmed to die in order for a normal worm to develop. A functional ced-9 gene [21], which is a Bcl-2 homolog, is essential for *C. elegans* to develop because it suppresses apoptosis in the cells that are normally destined to survive during development [20]. Mutations that inactivate ced-9 cause extra ectopic cell deaths to occur in the worm, whereas gain of function mutations of ced-9 prevent the deaths of the 131 cells that must die for normal worm development.

1.3
Suppression of Apoptosis in Mammalian Cells

In mammalian systems Bcl-2 expression correlates with the ultimate survival fate of many cell types during development. It is low in cells that are destined to die such as the inter-digital webs and in certain populations of thymocytes [24]. In the developing thymus Bcl-2 expression is low in the cortex, which consists of immature cells that are destined to die and is high in the thymic medulla which consists of mature thymocytes destined to become peripheral T cells [25]. Other experiments demonstrate that mice which are transgenic for expression of Bcl-2 in the T cell lineage have protracted survival in response to stimuli that kill normal T cells such as ligation of CD3 and deprivation of growth factors [26]. Mice that have a targeted disruption of IGF-I and the IGF-I receptor genes are 60% of normal size, have multiple abnormalities in cellular development, and die before birth, illustrating the importance of survival signals during normal mouse development [27].

Cell lines established in culture from different tissues are dependent on survival signals from surrounding cells or exogenously provided survival factors. In fact, there is now widespread acceptance for the hypothesis originally proposed by Raff [28] that all cells are dependent on survival signals, and without these signals they will undergo the default pathway to apoptosis. This dependence on survival factors provides organisms with a way to maintain tissue homeostasis. It also provides an innate defense against the development of tumors by allowing apoptosis to occur in cells in which mutations or de-regulated oncogenes arise.

Interestingly, a number of the survival factors now shown to be important suppressers of apoptosis have previously been well characterized as necessary growth factors for cells. These include IGF-I and IL-3 [18, 29]. Likewise, cytokines which were previously considered to be necessary for cell proliferation and to specify cell differentiation are now considered to be specific survival factors for cells as they undergo genetically pre-programmed differentiation [29–31].

2
Survival Factors in Cell Development

2.1
Survival Factors in the Nervous System

The importance and diversity of survival factors necessary for cell development, maintenance, and function is well illustrated in the central and peripheral nervous systems. The survival and maturation of developing neurons is dependent on the actions of neurotrophic factors secreted by the cells they innervate (Fig. 1). Neurons are generated in excess and approximately half of the neurons that are synthesized are eliminated by apoptosis [33–35]. When neurons reach their target tissue neurotrophins released by the target cells specifically suppress apoptosis in the exact number of neurons that are necessary to innervate the target tissue [36–38]. This competition for limited survival factors and the necessity for these factors to form high affinity interactions with their receptors regulates the numbers of neurons that survive and ensures that there is an exact numerical match between pre-synaptic neurons and post-synaptic target cells. In addition to their role in development, neurotrophins act as survival factors for sensory neurons for maintenance, and protect them in response to injury.

The neurotrophin family includes nerve growth factor (NGF), brain-derived growth factor (BDNF), neurotrophin-3 and neurotrophin-4. (NT-3 and NT-4/5/6) [39, 40]. All of the neurotrophin family share structural similarities and they interact with the Trk family of receptor tyrosine kinases and the low affinity p75 neurotrophin receptor [41–43]. NGF reacts with TrkA, BDNF with TrkB, and NT-3 with TrkC. TrkA and TrkB are also receptors for NT-3 and NT4/5, and all of the neurotrophins can interact with p75. During development TrkA has restricted expression in sensory and cholinergic neurons in the brain, whereas TrkB and TrkC are more widely expressed in the CNS [44]. P75 is expressed in

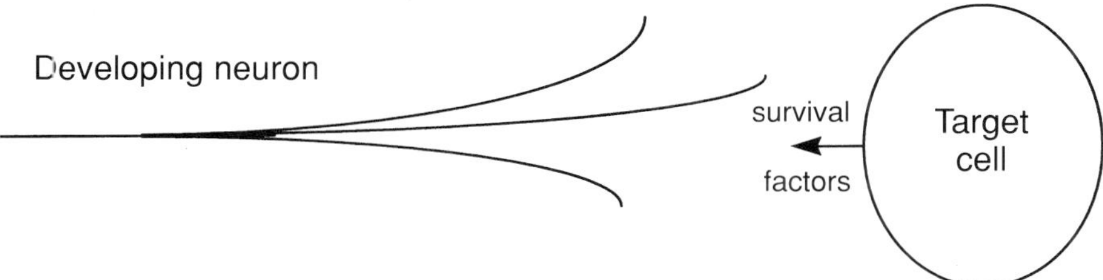

Fig. 1. Developing neurons are dependent on survival factors released by the cells they innervate. Survival factors for different sub-groups of neurons include the neurotrophins, NGF, BDNF, NT-3/4/5; IGF-I; FGFs; and TGFs. The neurotrophins act through the Trk family of receptors expressed on neurons. IGF-I acts through the IGF-IR, and the FGFs and TGFs act through their cognate receptors

many cell types including motor neurons and Schwann cells [45, 46]. Downstream signaling from neurotrophins involves activity of PI-3 kinase, which is discussed in more detail below [47].

Like sensory neurons, the development of motoneurons in the spinal cord is dependent on their interactions with the tissue they innervate: skeketal muscle and other central nervous system cells. Survival factors for these neurons include members of the FGF (fibroblast growth factor) family and TGF-beta (transforming growth factor) family, CNTF (ciliary neurotrophic factor), and CDF-LIF (cholinergic development factor-leukemia inhibitory factor) [48]. In addition IGF-I has been shown to be a survival factor for rat motor neurons [49] and the neurotrophins NT-4/5 prevent injury-induced death in facial motor neurons in neonatal rats [50].

Cerebellar granule neurons are dependent on high levels of K+ for their survival and development, but are very responsive to IGF-I as a survival factor when they are placed in low K+ concentrations [51].

Midbrain dopaminergic neurons, which regulate motor activity, emotional behavior, and cognition, are also kept alive by survival factors such as GDNF, (glial cell derived neurotrophic factor), TGF beta 2 and TGF beta 3, in culture [52] These survival factors are expressed in the target tissues of dopaminergic neurons.

Non-neural cells important for the function of the nervous system include those derived from glial cells. In the developing rat optic nerve approximately half of the newly formed oligodendrocytes die due to competition for survival factors. Both PDGF and IGF-I are survival factors for these cells and their precursors in culture [53]. The PDGF receptor is highly expressed on newly-formed oligodendrocytes and their progenitors. It is thought that the competition for survival factors in these cell lineages contributes to match the number of oligodendrocytes to the number of axons needing myelination [53].

The PC12 rat phenochromocytoma cell line is used as a model of neuronal cell survival, and interestingly NGF, EGF, and IGF-I can all suppress apoptosis induced by serum withdrawal in these cells [47, 54].

2.2
Eye Development

Survival factors are critical for the development and maintenance of cells in the eye. Survival factors produced by lens epithelium cells are important for development of neighboring epithelial cells in the lens of the eye [55]. In the retina [56], basic FGF can protect the photoreceptors in rat retinas from light damage and there is evidence that retinal injury stimulates intrinsic survival mechanisms [57].

2.3
Development of Gonads

Normal development of male and female gonads is achieved through hormonal regulation of apoptosis. In the ovary more than 99% of follicles selected to mature will die in a lifetime and in the testis up to 75% of the male germ cells undergo apoptosis [58]. In the ovary gonadotropins, estrogens, growth hormone, IGF-I, EGF, TGF alpha, basic FGF, and IL-1 are involved in maintaining the survival of pre-ovulatory follicles [58]. Follicle stimulating hormone can protect these pre-ovulatory early antral follicles from apoptosis in culture [59]. These follicles are dependent on survival factors in order to mature and secrete estrogen to trigger ovulation.

In the testis, androgens and gonadotropins are necessary survival factors, and androgens are also survival factors for both normal and malignant prostate cells [60].

3
Role of Cytokines in Survival and Differentiation of Hemopoietic Cells

The expansion, differentiation, and maintenance of hemopoietic cells has long been known to be dependent on soluble factors or cytokines. IL-3, GM-CSF, stem cell factor (SCF), and erythropoietin (Epo) were shown to be required for the maintenance, proliferation, and differentiation of bone marrow progenitors in culture [61]. Many cytokines were later shown to inhibit specifically apoptosis in bone marrow and hemopoietic cell lines without inducing proliferation [32, 62–65]. It has been suggested that the role of cytokines in hemopoietic cell differentiation, rather than being instructive in specifying lineage choice, may be to suppress apoptosis in certain populations of cells (Fig. 2). This has been inferred from the observation in a multipotential hemopoietic cell line that Bcl-2 promotes survival and multi-lineage differentiation in the absence of exogenously added cytokines [30].

A similar conclusion can be reached from studies of the Epo receptor and its signaling pathway that have been informative in delineating the role of this cytokine in the proliferation, differentiation, and survival of cells of the erythroid lineage. Mice which had both Epo and the Epo receptor knocked out (homozygotes) died prematurely at embryonic day 13, and the mice demonstrated reduced liver erythropoiesis [66]. Interestingly, the mouse embryos had BFU-E and

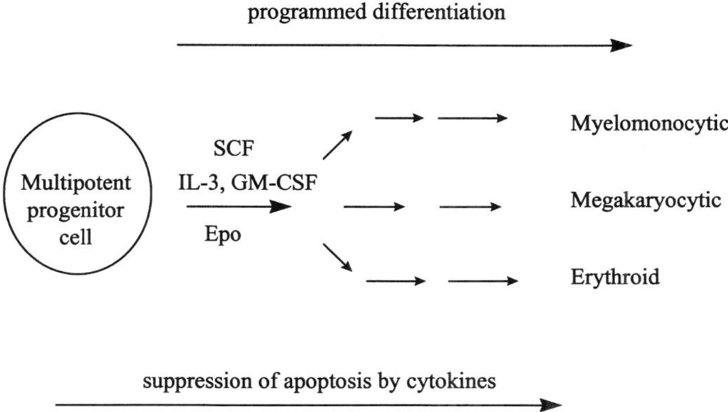

Fig. 2. Cytokines act to suppress apoptosis during hemopoietic cell differentiation. Hemopoietic progenitor have a pre-programmed differentiation pathway. Cytokines such as IL-3 and Epo are thought to facilitate this process by suppressing apoptosis in selected populations of progenitor cells

CFU-E present in their fetal livers, thus demonstrating that Epo and the EpoR are not required for lineage commitment or for the differentiation of BFU-E to CFU-E progenitors, but rather are required for the survival and proliferation of the CFU-E progenitors and their subsequent irreversible terminal differentiation.

By generating chimeric receptors with a truncated EpoR or wild type EpoR and the prolactin receptor followed by transfection of these chimaeric receptors into primary erythroid progenitors, it was shown that cells expressing either the truncated (membrane-proximal 136 amino acids) or wild type cytoplasmic domain of the EpoR fused to the extracellular domain of the prolactin receptor could proliferate and form erythroid colonies in response to the hormone prolactin [67]. This demonstrated that the same minimal region of the EpoR is sufficient to support proliferation and differentiation, suggesting that the deleted domains of the receptor are required for the survival activity of the receptor.

Hemopoietic cells can be protected by IGF-I from cell death induced by cytokine withdrawal and other stimuli, suggesting that this growth factor may also play a role in the survival of these cells during development (discussed below). The survival factors that may be involved in keeping terminally differentiated blood cells alive are less well understood. IL-4 and IL-2 have been shown to rescue selectively Th 1 and Th 2 subsets of T cells, respectively, from glucocorticoid-induced apoptosis [68]. However, whether mature B cells, monocytes, or terminally differentiated granulocytes are dependent on specific survival factors is not known.

4
The IGF-I/IGF-IR System in Cell Survival

4.1
Ligands

IGF-I and IGF-II are single chain peptides of approximately 70 and 67 amino acids, respectively, which were initially characterized as having mitogenic actions as well as insulin-like activities in adipose and muscle tissue [69]. The IGFs have overall approximately 40–50% homology between themselves and insulin [70]. The single chain peptides consist of four peptide domains, B, A, C, and D, where the domains A and B are homologs of the A and B domains in insulin. The C domain is homologous to the connecting C peptide that is cleaved out of pro-insulin, whereas the D domain is not found in insulin. The IGFs also include E peptides that are cleaved during processing [71]. Folding of IGF-I and IGF-II is stabilized by three intrachain disulfide bonds. The IGF-I gene is mapped to the long arm of chromosome 12 and consists of at least six exons, whereas IGF-II is located on the distal end of the short arm of chromosome 11, contiguous to the insulin gene and includes nine exons [72]. The transcription of both IGF-I and IGF-II is very complex with multiple leader exons controlled by several promoters. Expression of the IGF-I and IGF-II genes is developmentally regulated with IGF-I levels increasing by 10- to 100-fold in all tissues between birth and adulthood [73], whereas IGF-II levels are high before birth and decline afterwards in rat, but remain present at detectable levels in human serum. Studies with knock out mice where the IGF-I or IGF-II genes were targeted by homologous recombination revealed that the mice were born at 60% of normal birth weight, and many of them die shortly after birth [27]. These mice demonstrated underdeveloped bone, muscle, and lung. This indicates that these genes play a critical role in embryonic growth and differentiation.

IGF-I is also known as somatomedin C, due to it being identified as the mediator of growth hormone action [74] in the so-called somatomedin hypothesis. IGF-I is a specific regulator of growth hormone gene transcription, mRNA levels, and protein secretion in pituitary cells. These cells also express the IGF-I gene, which is regulated by growth hormone and suggests that IGF-I plays a paracrine role in pituitary cell function [75]. Unlike insulin, the IGFs are ubiquitously expressed, being produced in large quantities by the liver and lesser quantities by other organs. Generally IGF-I and IGF-II are thought to act both as endocrine hormones via the blood and as paracrine and autocrine growth factors locally [70]. IGF-I and IGF-II are also thought to play an important role in the development and/or function of the central nervous system, where 50 ng/ml IGF-I and 2 ng/ml IGF-II have been measured [76]. Apparently, there is differential expression of IGF-I and II in different regions of the brain, different isoforms of each are present, and these proteins are thought not to be regulated by growth hormone (reviewed in [70, 76]).

IGF-I and IGF-II are found circulating in serum at high concentrations (100nmol/l),. This is 100 times higher than the concentration of insulin. However, the IGFs are present in association with high affinity IGF binding proteins

(IGFBPs), which are thought to sequester them away from insulin receptors, facilitate their transport, and regulate their biological functions at a cellular level. At least six of these binding proteins have been identified and cloned [77]. For example IGFBP3 has been shown to sequester free IGF-I, and block the growth promoting action of IGF, but more recently it has been shown to mediate induction of apoptosis and function independently of IGF or its receptor [78]. Although the IGFs and insulin share significant homology and can cross react with the heterologous receptors, they do have unique binding sites. Insulin cannot interact with the IGFBPs, and analysis of these cross reactions has allowed for definition of the key residues required for binding by the different peptides [70]. IGF-I and IGF-II interact with the type IGF-I receptor as well as the insulin receptor. Insulin can also interact with the IGF-I receptor and the affinity of binding of IGF-II and insulin to this receptor are 10-fold and 100-fold lower, respectively. IGF-II also binds to the IGF-II receptor, also known as the mannose 6-phosphate receptor [79].

4.2
IGF-I as a Survival Factor

IGF-I has been recognized as an important mitogen for many cell types for a considerable period of time. For example, fibroblasts usually require IGF-I and platelet-derived growth factor (PDGF) or epidermal growth factor (EGF) to grow optimally (reviewed in [80]). These growth factors are normally provided by fetal bovine serum (FBS) used in cell culture. However, over-expression of the IGF-IR in fibroblasts results in cells that require only IGF-I and no other growth factors to proliferate [81]. In fact a number of oncogenes and growth factors have been shown to increase the expression of IGF-I and thereby contribute to the growth of tumor cells [82].

The role of IGF-I as a survival factor for cells in culture was first demonstrated by Rodriguez-Tarduchy et al., using IL-3-dependent cell [83]. The BAF-3 cell line is an IL-3-dependent pre-B cell line and the FDCP-Mix cell line is derived from a long-term bone marrow culture. Both require IL-3 for their propagation in culture and removal of IL-3 results in cell death with all the hallmarks of apoptosis. This cell death could be blocked in both cell lines by adding IGF-I to the cultures. IGF-I was only very weakly mitogenic to these cells compared with IL-3, thereby suggesting that it had a specific survival activity for hemopoietic cells. Indeed the establishment and maintenance of cytokine-dependent cell lines from leukemic bone marrow requires culture conditions that includes the specific cytokine and medium supplemented with either FBS or high concentrations of insulin [84, 85]. It is likely that these concentrations of FBS or insulin are sufficient to activate the IGF-IR, and indeed all cytokine-dependent cell lines tested to date can be protected from apoptosis induced by cytokine withdrawal with IGF-I (unpublished observations). This supports the idea that IGF-I is generally required for hemopoietic cell survival and that it is required for most cell types to grow in culture [86].

The role of IGF-I as a survival factor was firmly established by the study of c-Myc induced apoptosis in Rat-1 fibroblasts cultured in low serum [18]. In this system, Rat-1 fibroblasts expressing an eostrogen-regulatable c-Myc fused to the estrogen receptor undergo apoptosis when placed in low serum culture conditions [87]. The cells continue to proliferate under the influence of c-Myc in both the cultures that are supplemented with 10% FBS and those in low serum, but the culture with low serum loses viability. This lead to the search for a survival factor present in serum that would protect Rat-1 cells from c-Myc-induced death. IGF-I proved to be the most potent protector from c-Myc induced death, followed by IGF-II and PDGF, which were all better survival factors than known mitogens for these cells EGF and FGF. Interestingly, IGF-I could protect from death at any stage of the cell cycle and even in the presence of the protein synthesis inhibitor cycloheximide [18]. The level of survival activity provided by IGF-I in this system was similar to that provided by expression of the anti-death gene Bcl-2 [88]. This all suggested that IGF-I survival activity was distinct from its mitogenic activity and that it was an epigenetic event, not requiring new protein or gene synthesis.

4.3
IGF-I Protects a Wide Range of Cells from Apoptosis

In addition to protecting hemopoietic cells and fibroblasts from apoptosis, IGF-I has been shown to be anti-apoptotic in many other cell systems. Several studies have been done in cells of the central and peripheral nervous system, both in vitro and in vivo. In oligodendrocytes and their precursor glial cells cultured from the rat optic nerve (where 50% of newly formed oligodendrocytes die by apoptosis), both IGF-I and PDGF have been shown to be potent survival factors [89]. In rat motor neurons during development and adulthood, IGF-I has been shown to mediate survival activity in response to damage or to neurotrophic withdrawal [49]. Cerebellar granule neurons in culture can be protected from apoptosis by IGF-I [51] and neuroblastoma cells can be protected from hyperosmotic stress by IGF-I [90]. IGF-I also protects neuronal cells from injury such that induced by ischemia and elevated calcium [91, 92]. In fact IGF-I levels have been shown to increase following nerve damage in vivo, suggesting that IGF-I functions as an endogenous neuro-protectant [93]. IGF-I has been approved as a therapeutic agent for the neurodegenerative disease chronic amylolateral sclerosis (Lou Gehrigs disease), which is characterized by rapid progressive loss of neuromuscular function [94].

IGF-I and IGF-II are potent survival factors for a wide range of tumor cells, in particular those derived from breast such as MCF-7 and lung, when induced to undergo apoptosis by serum withdrawal or chemotherapeutic drugs (unpublished results). These cytokines are often produced by cell lines and found at elevated levels in the circulation of tumor patients (reviewed in [95, 96]). There is abundant evidence to indicate that IGF-I and IGF-II are critical survival factors for the development and progression of carcinogenesis, and that tumors are indeed dependent on these survival signals. For example, mice that have the IGF-II genes knocked out develop cancers at less frequency than normal mice and the tumors that arise are less malignant than tumors that arise in normal mice [97].

The association of increased IGF-II with cancer was documented by analysis of gene expression levels in normal cells vs tumor cells [98]. Using a newly developed method called serial analysis of gene expression (SAGE), which allows for the rapid analysis of the expression levels of thousands of genes, the authors quantified the copy number of transcripts for a total of 49,000 genes in normal colorectal epithelium, colorectal cancers, or pancreatic cancers. The transcript for IGF-II had the greatest increase in expression in colorectal cancers out of a total of 289 transcripts that were increased relative to normal cells [98]. The role of the IGF-I family as survival factors in tumors is discussed in more detail below in the section on IGF-IR in tumors.

4.4
IGF-I Protects Cells from Diverse Death Stimuli

In addition to protecting cells from cytokine and serum withdrawal, IGF-I and IGF-II have been shown to protect different cells from diverse cell death-inducing stimuli. All of the anti-apoptotic activity is mediated through the type I IGF-I receptor. Cells can be protected from chemotherapeutic drugs such as etoposide by IGF-I [99], and from ultraviolet irradiation [100]. Over-expression of the death promoting members of the Bcl-2 family Bak or Bax leads to extensive apoptosis in IL-3-dependent cells such as FL5.12 cell and in fibroblasts [17]. IGF-I can protect Bak or Bax-expressing Rat-1 cells from death in low serum [17], and can protect FL5.12 cells which co-express the IGF-IR with Bak or Bax from cell death (unpublished results). IGF-I can also protect cells from the pro-apoptotic effects of activating pro-ice to ICE (caspase I) by proteolytic cleavage [101], a key step in initiating apoptosis in all cells. The fact that IGF-I can protect from such a wide variety of pro-apoptotic signals and from the effects of expressing key regulators of the apoptotic pathway such as Bcl-2 family members and caspase-1 indicates that it can intersect at a key stage of the apoptosis regulatory mechanisms that determine the life or death fate of a cell in response to signals from its environment.

5
The IGF-I Receptor

5.1
Structure and Properties

The IGFs mediate their actions by binding to a family of trans-membrane receptors which include the type I IGF-I receptor, the IGF-II/mannose 6 phosphate receptor, and the insulin receptor (reviewed in [95]). In the context of cell survival, only the type I receptor (IGF-IR) will be discussed. This receptor can bind IGF-I with an affinity of 1 nmol/l, IGF-II with a 2–50 fold lower affinity, and insulin with a 100–500-fold lower affinity [70]. However, all three ligands can mediate the same functions via the IGF-I receptor.

The IGF-IR is encoded by a 100 kb gene containing 20 exons located on the distal arm of chromosome 15 [102]. The molecular organization of the gene and the primary sequence shares significant similarity to the insulin receptor. It has a very large 5′ untranslated region and transcription has shown it to be regulated by several transcription factors associated with mediating cell surface signals, including members of the EGR family such as Spi 1. Tumor suppresser genes such as the Wilms tumor gene WTI [96, 103] and p53 [104] have also been shown to repress expression of the IGF-IR. The inactivation of WTI or p53 along with the resultant increase in IGF-IR expression has been implicated in the etiology and progression of Wilms tumor (a common childhood renal cancer) and other cancers.

Transcripts of IGF-IR mRNA are low in abundance but can be found in almost all tissues and cell types, except for liver, the major site of IGF-I [105, 106]. Studies in rats indicate that levels of receptor expression tend to be highest in the embryo and they decrease in adult animals, with the central nervous system having the highest levels followed by kidney, testes, lung, stomach, and heart [106].

The IGF-IR gene encodes a single precursor protein consisting of 1367 amino acids [107] and containing a 30 residue signal peptide. It is then cleaved at residues 707–710 to form alpha and beta subunits which are stabilized by primary disulfide bonds. The IGF-1R forms a tetrameric protein consisting of two alpha chains and two beta chains held together by secondary disulfide bonds (Fig. 3). The alpha chain resides outside the cell and contains a cysteine-rich motif constituting the ligand binding site. The beta chain contains a hydrophobic transmembrane domain and a kinase domain as well as several tyrosine autophosphorylation sites. These include the tyrosine cluster in the kinase domain at Y1131, 1135, 1136, phosphorylation of which is critical for kinase activity. Y950 in the juxtamembrane region flanked by the NPEY motif is required for binding the PTP domains in the IRS-1 protein or SHC [108]. Y1316F in the C-terminal tail of the IGF-IR is flanked by the consensus sequence for binding of the SH2 domain in the p85 subunit of PI-3 kinase [109].

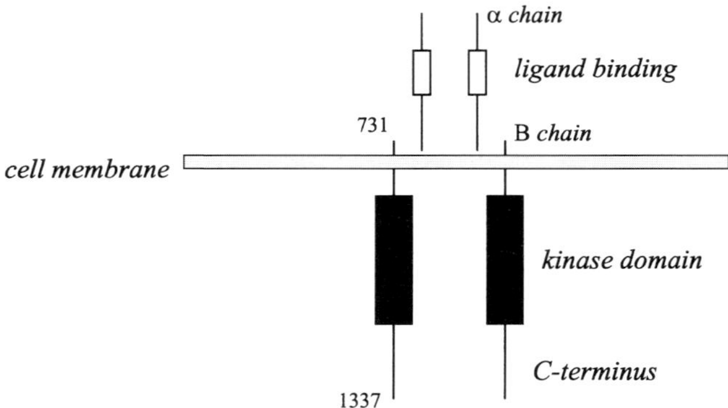

Fig. 3. Schematic diagram of the IGF-IR. This is a tyrosine kinase receptor consisting of two alpha chains and two B chains that are stabilized by disulfide bonds. It can bind the ligands IGF-I, IGF-II, or insulin at a cysteine-rich region in the alpha chain. The B chain has a tyrosine kinase domain and a C-terminal region Amino acid numbering is according to Ullrich et al. [107]

Overall the IGF-IR shares significant homology with the insulin receptor, particularly in the kinase domain, which has 80–90% amino acid identity to the insulin receptor kinase domain. The tyrosine clusters are conserved as well as tyrosines in equivalent positions to Y950 and Y1316. The C-terminus of the IGF-IR shares less homology with the insulin receptor than other regions of the IGF-IR, with overall 40% identity. There are two tyrosines at positions 1250 and 1251 that are not present at the equivalent position as well as a stretch of basic amino acids from residues 1293 to 1301 that is not present in the insulin receptor. These regions will be discussed in more detail in section in the section which discusses the mutational analysis of the IGF-IR. Conversely, there is tyrosine present in the insulin receptor that is not present in the IGF-IR.

There are reports of IGF-IR alpha and beta chain variation in different tissues [95]. Beta chains with molecular weights of 95 or 105 kDa have been detected, whereas the cloned IGF-IR cDNA encodes for a 105 kDa protein. It has been suggested that the cloned protein may be the fetal IGF-IR, whereas the 105 kDa protein is the adult IGF-IR beta chain. The existence of any functional differences between the different molecular weight receptors is unknown.

5.2
The IGF-IR in Normal Growth and Development

The expression levels and the survival activity of the IGF-IR all point to it having a very important role in normal growth and development. This was underscored by knock out mouse studies where it was shown to be an essential gene [27]. The homozygous progeny of mice which had a targeted disruption of the IGF-IR gene (IGF-IR–/–)were only 45% the size of normal mice and died at birth. Examination of these mice revealed that they had impaired lung function and could not breathe. In addition, examination of the IGF-IR KO mice embryos demonstrated that they had muscle hypoplasia, reduced numbers of oligodendrocyte precursors in the central nervous system, thinner epidermal layers, and development delays in ossification of bones of the trunk and extremities. Overall this study combined with the studies on IGF-I and IGF-II knock out mice indicates that the IGF-IR is essential for normal embryonic development.

Studies in cultured cell lines indicate that IGF-IR activity is essential for cells to traverse the cell cycle (reviewed in [80]). Antisense strategies which reduce IGF-IR numbers block the growth of several cell lines and it has been proposed that the growth stimulating activities of EGF and PDGF are mediated through the IGF-IR [80].

5.3
Survival Signals from the IGF-IR are Essential for Tumorigenesis and Maintenance of Tumor Cells

The establishment and maintenance of tumors is dependent on the availability of survival signals to suppress apoptosis, and one critical survival signal for tumors is provided by activation of the IGF-IR by either IGF-I or IGF-II. There are several lines of evidence to support this view.

One line of evidence is provided by elegant studies in a mouse model system where the simian virus-40 large T-antigen (Tag) is targeted to the islets of Langerhans under the control of the insulin gene regulatory region, resulting in the development of hyperplasia and tumors in the islets [110, 97]. In this system IGF-II expression was up-regulated in the tumors concomitant with the onset of hyperplasia, and the development and maintenance of tumors was dependent on the availability of IGF-II. Interference with IGF-II expression by antisense strategies reduced the proliferation of tumor cells in vitro. Strikingly, Rip-Tag transgenic mice developed on an IGF-II null background developed tumors with reduced malignancy, and although they had similar mitotic indices to IGF-II wild-type tumors, they exhibited an increased apoptotic index resulting in significantly smaller tumors. This indicates that IGF-II provides a critical survival signal for the development of these tumors. Expression levels of the Bcl-2 family death suppresser gene Bcl-xl was up-regulated in the RIP-Tag tumors, and transfection of Bcl-xl into the tumor cells resulted in the development and enlargement of the RIP-Tag tumors due to suppression of apoptosis in the tumor cells [111] The RIP-Tag model illustrates that the Tag expressing cells, which are pushed to proliferate, are dependent on survival signals in order to evolve into a tumor.

The cooperation of oncogenes in the process of tumorigenesis is well documented. For example, transgenic mice expressing c-Myc and Bcl-2 develop tumors more rapidly than mice expressing Myc alone [112, 113]. In an in vitro system IGF-I can protect rat-1 fibroblasts from c-Myc-induced death [17] as effectively as Bcl-2 [88]. This observation, taken together with the role of IGF-II and Bcl-xl in the RIP-Tag system, indicates that the crucial cooperating signal provided by Bcl-2 family members or by IGF-1 is a survival signal that suppresses apoptosis. Therefore, one would predict that over-expression of the IGF-IR would be a cooperating oncogene along with another oncogene that drives proliferation such as c-Myc. This is illustrated in Fig. 4.

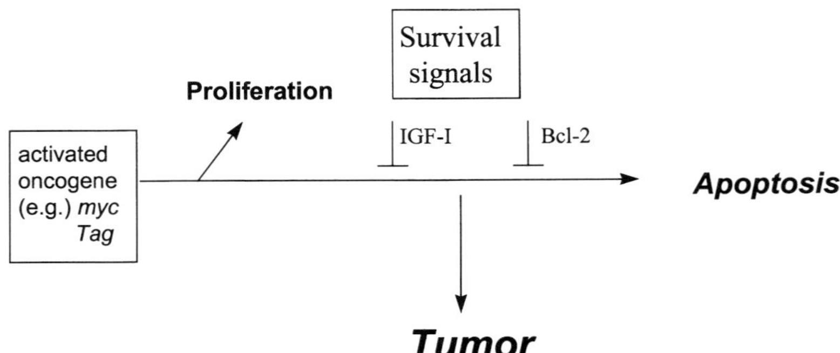

Fig. 4. Model for role of survival signals in tumorigenesis. Cells expressing an activated oncogene, which drives them to proliferate, would normally undergo apoptosis. However, the availability of survival signals, either from up-regulated expression of the IGF-IR and ligands, or from an anti-death gene such as Bcl-2, allows the oncogene-expressing cells to survive and evolve into a tumor

The IGF-IR has long been documented to have a role cellular transformation. Over-expression of the IGF-IR transforms fibroblasts and allows them to grow in an anchorage-independent manner in soft agarose or from tumors in nude mice [114]. Transformation of cells by the SRC oncogene of Rous sarcoma virus (pp60src) results in constitutive phosphorylation of the IGF-IR [115]. Cells derived from the IGF-IR knock out mouse are refractory to transformation by a series of oncogenes and growth factor receptors including SV40 Tag and Ha-ras, and the PDGF receptor [80, 116]. Susceptibility to transformation can be restored by reintroducing the IGF-IR. The transforming signal from the IGF-IR is distinct from a mitogenic signal because receptors that have the C-terminus deleted lose the ability to transform cells but retain mitogenic capability [117].

5.4
Abrogation of IGF-IR Expression and Function Leads to Apoptosis in Tumor Cells

The IGF-IR is over-expressed on many tumor cells as determined by binding studies with fresh biopsies and in vitro cultured cells (reviewed in [95, 96]). These include tumors of the breast, lung, pancreas, glioblastomas, and others. A number of approaches have been undertaken to investigate the effects of inhibiting IGF-IR expression and function in tumors. These include inhibitory antibodies such as the anti-IR-3 antibody [118], peptide analogs of IGF-I [119], and antisense strategies against the receptor [120–123]. In breast cancer lines the anti-IR3 antibody could inhibit anchorage-independent growth in the presence of serum and inhibited the formation of tumors in nude mice [118]. However, the antibody was not effective in blocking growth in serum-free medium in vitro or against established tumors in nude mice. Similarly, peptide analogs could inhibit the growth of prostatic cancer cell lines in vitro [119]. Dominant negative receptors gave different results depending on the characteristics of the dominant-negative construct used. A receptor that was truncated at residue 952 and presumably could not cross-phosphorylate endogenous receptors or be phosphorylated by endogenous receptors was effective in inhibiting tumorigenesis in vivo [124]. However intact receptors with point mutations in the kinase domain or at tyrosines in the β chain could inhibit growth of C6 gliobalstoma cells but not inhibit tumorigenesis [125]. These receptors could probably form chimeras with endogenous receptors and become trans-activated functional receptors.

Expression of antisense oligonucleotides or RNA for the IGF-IR leads to reversal of the transformed phenotype and apoptosis of tumor cells [120–123] The effects of antisense oligonucleotides was studied on several cell lines including C6 rat glioblastoma cells, human melanoma cells, and mouse leukemia cells by measuring colony formation in soft agarose, survival in a diffusion chamber in vivo, and the formation of tumors in nude mice. [120, 121]. It was observed that the key requirement in reversing the transformed phenotype was to reduce the numbers of functional IGF-IRs on the tumor cells. A 60–70% decrease in IGF-IR numbers on the cells was sufficient to inhibit colony formation, induce extensive apoptosis in the diffusion chambers in vivo, and prevent tumor formation in nude mice. This result supports the hypothesis that tumor cells are dependent on suppression of apoptosis by survival signals for their maintenance, and disruption of these survival signals leads to apoptosis and the demise of the tumor. This raises the intriguing question whether the transformation signal provided by the IGF-IR is in effect a survival signal? This question is addressed in the next section on the mutational analysis of the IGF-IR.

6
Mutational Analysis of the IGF-IR

The IGF-IR mediates suppression of apoptosis, mitogenesis, and transformation. The signals from the receptor for mitogenesis and transformation are distinct because receptors that have the C-terminus deleted lose the ability to transform cells [117]. The fact that suppression of apoptosis could take place without new gene or protein synthesis suggested that suppression of apoptosis was a distinct signal from mitogenic signaling [18]. However, the role of the IGF-IR in the progression of tumorigenesis [111] and the fact that the IGF-IR is important for the maintenance of the transformed phenotype [120, 121] suggested that the signals from the receptor for transformation may be overlapping with those mediating suppression of apoptosis. Therefore, in order to assign domains to the three different signals that originate from the receptor, a mutational analysis of the IGF-IR was undertaken [128]. We wished to determine if there was a domain in the receptor that was essential for the anti-apoptotic or survival activity of the receptor, and to determine if the domain requirements for the anti-apoptotic and transforming activities were over-lapping.

6.1
Transformation and Mitogenic Activity of IGF-IR Mutants

A series of mutants of the IGF-IR had been previously generated by Baserge et al., and analyzed for their mitogenic and transforming activity in fibroblasts derived from the IGF-IR null mice (R-cells) (summarized in [126]). R-cells cannot be transformed by a number of oncogenes or growth factor receptors until IGF-IR expression is restored. However, over-expression of the IGF-IR is sufficient to transform the cells [116]. The IGF-IR mutants studied included receptors which had the kinase domain inactivated by mutation of the critical ATP binding lysine 1003 to arginine; key tyrosines switched to phenylalanine, such as Y950 (the IRS-I binding site); the tyrosine cluster in the kinase domain at Y1131, 1135; and 1136, and Y1136, which is a P-I3 kinase binding site.

A series of IGF-IR mutants that had alterations in the C-terminus were also tested for transforming and mitogenic activity. These receptors were mutated at residues that are unique to the IGF-IR and are not present at equivalent positions in the insulin receptor. These residues were tyrosine 1250, 1251; four serines at 1280–1283; and histidine 1293/lysine 1294. In R- cells the C-terminal region of the IGF-IR was shown to be essential for transformation [117], and receptors which were truncated to amino acid 1229 by deletion of the entire C-terminus failed to transform R⁻ fibroblasts, but retained full mitogenic potential in response to IGF-I. Within the C-terminal region, the transforming activity was further localized to a domain between amino acids 1245 and 1310 [126].

However, all of the C-terminus point mutant and truncated receptors retained mitogenic capability. These studies indicated that the mitogenic and transforming functions of the IGF-I receptor were mediated by spatially distinct domains. Mutations at the kinase domain ATP binding site that ablate kinase activity, at the tyrosine cluster in the kinase domain, or at tyrosine, abolished both proliferation and transformation [127], which demonstrates that these residues were required for both mitogenic and transformation signaling.

6.2
IGF-IR Mutants and Suppression of Apoptosis

The same group of IGF-IR mutants used to map receptor domains involved in transformation and mitogenesis were tested for their abilities to mediate cell survival in two cell systems [128]. Analysis of IGF-IR-mediated suppression of apoptosis was complicated by the fact that R- cells were not suitable for studying apoptosis because over-expression of the IGF-IR lead to ligand-independent survival, probably due to autocrine mechanisms. Therefore two different cell systems were used [128]. One was cell death induced by IL-3-deprivation in the murine FL5.12 B lymphoblastoma cell line. FL5.12 cells are dependent on IL-3 for both survival and growth and express low numbers of endogenous IGF-IR. However, FL5.12 cells transfected with WT IGF-IR were protected by FBS and IGF-I from apoptosis following IL-3 withdrawal, which indicates that IGF-IR could activate a pre-existing survival signaling pathway in these cells (Fig. 5). Parallel studies with some of the mutants were carried out in Rat-1 fibroblasts although these cells express a significant level of endogenous IGF-I and IGF-IR that could complicate the signaling from ectopically expressed receptors.

A summary of the locations of the various mutants and the results of the analysis of suppression of apoptosis in FL5.12 cells is shown in Fig. 6. In both cell models inactivation of the critical ATP-binding lysine reside K10003 completely ablated survival, indicating the necessity for kinase activity for suppression of apoptosis. In contrast, mutation of tyrosine 950, a residue involved in interaction with the IRS-1 "docking protein" and SHC or of the tyrosine cluster (Y1131, Y1135, Y1136), had little discernible effect on IGF-IR-mediated suppression of apoptosis in FL5.12 cells and Rat-1/Myc cells. Neither of these mutants had mitogenic and transforming functions in IGF-IR null cells [126].

Two IGF-IR C-terminal point mutants Y1251F and H1293F/K1294R, abolished IGF-I-mediated protection from IL-3 withdrawal (see Fig. 6). These mutants were also inactive in transformation, although they retained mitogenic activity when expressed in IGF-IR null fibroblasts [126]. The S1280–1283 A IGF-IR mutant was similarly active mitogenically and inactive in transformation in IGF-IR null fibroblasts. However, in contrast to the Y1251F and H1293F/K1294R mutants, mutant S1280–1283 A retained the ability to elicit a survival signal in FL5.12 cells. IGF-IR mutants in which the entire C-terminus of the receptor was truncated, for example mutants d1229 and d1245, retained both their anti-apoptotic and mitogenic [128] activities, although they were inactive in transformation [126]. A comparison between the activity of the IGF-IR mutants in suppression of apoptosis and transformation is shown in Table 1. Taken together,

FL5.12/neo

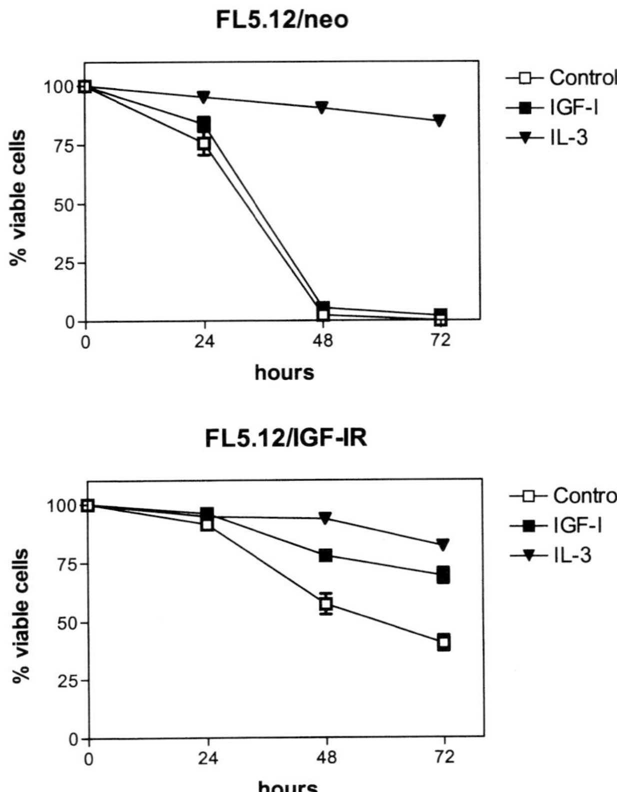

Fig. 5. IGF-I protects FL512/IGF-IR cells from apoptosis due to IL-3 withdrawal. FL512 cells expressing the WT IGF-IR or a control vector (neo) were cultured for 24 h in the presence of IL-3, washed extensively, and cultured in medium containing 5% FBS (control), 5% FBS + IGF-I, or 5% FBS + IL-3. Cell viability was monitored over 72 h by trypan blue exclusion, and the data represent the mean and standard deviation of percent viability in triplicate cultures, plotted as a function of time. FL512 cells expressing the various mutant receptors described in Table 1 and Fig. 6 were assayed in a similar manner and results compared to WT and neo cells

these data indicated a degree of overlap between the transforming and anti-apoptotic functions of the receptor with regard to their requirement for certain residues in the C-terminal region of the receptor. Suppression of apoptosis appeared to be necessary for IGF-IR-mediated cell transformation, but was not sufficient. Whereas all mutants that were inactive in suppressing apoptosis were also inactive for cell transformation, some mutants that were inactive in transformation remained active in suppressing apoptosis. Thus, it appeared that in addition to the suppression of apoptosis, transformation required additional signals arising from the IGF-IR. These signals could possibly be mediated by the serine residues at 1280–1283 which are essential for transformation.

The activity of the IGF-IR mutants in Rat-1 cells and FL5.12 cells also confirmed that the regions of the IGF-IR required for inhibition of apoptosis and for

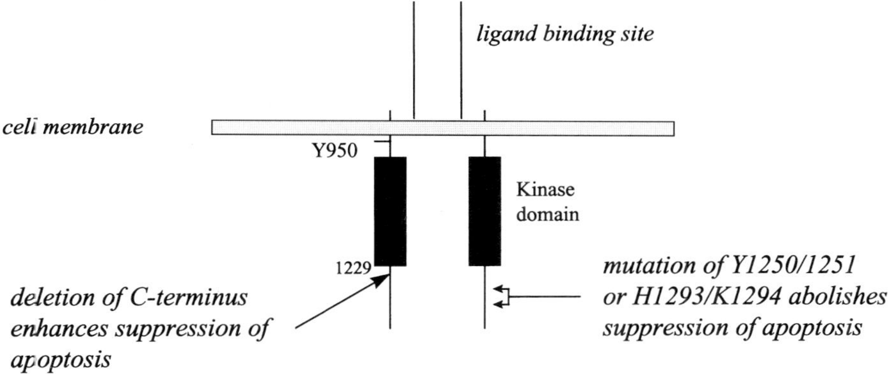

Fig. 6. Summary of domains of IGF-IR involved in suppression of apoptosis. IGF-IR mutants were assayed for suppression of apoptosis as outlined in legend to Fig. 5. Receptors mutated at Y950 retained the ability to suppress apoptosis, whereas mutants Y1250/1251F and H1293F/K1294L lost the ability to suppress apoptosis Truncated receptors which had the entire C-terminus removed had enhanced suppression of apoptosis

Table 1. Summary of nitogenic, transforming, and anti-apoptotic function of IGF-IR mutants

Receptor	Interaacting protein, or function of mutated residues	Mitogenic activity	Transforming activity	Suppression of apoptosis
wild type		yes	yes	yes
Y950F	IRS-1, SHC	no	no	yes
K1003A	ATP binding			
Y1131, 1135, 1136F	tyrosine cluster (kinase domain)	no	no	yes
Y1250F	?	yes	yes	yes
Y1251F	?	yes	no	no
S1280 – 12283A	?	yes	no	yes
H1293F/K1294L	?	yes	no	no
Y1316F	p85 of PI-3 kinase	yes	yes	yes
d1229 deleted at 1229		yes	no	yes

[a] Data with R⁻ cells derived from references (116, 126, 127, 128).
[b] Data summarized from IGF-1-mediated protection afforded by mutant IGF-IRs in IL-3 withdrawal assays with FL 5.12 cells (128).

mitogenesis are distinct. Mutants such as Y950F and Y1131, 1135, 1136F were inactive for mitogenesis yet retained anti-apoptotic activity. Conversely, mutants such as Y1250F/Y1251F and H1293F/K1294R could not promote survival yet they retained mitogenic capacity.

One interesting but somewhat paradoxical result came from the mutational analysis of the IGF-IR. The IGF-IR C-terminal truncation mutants d1229 and

d1245 retained anti-apoptotic activity, while point mutations within the same
C-terminal region, such as Y1251F and H1293F/K1294R, ablated it. In fact the
truncated receptors had enhanced anti-apoptotic activity and were constitutively active for survival in Rat-1/myc cells. There are two possible
explanations for this. One is that the C-terminal domain of IGF-IR has intrinsic pro-apoptotic activity, although analysis of the sequence does not reveal
any similarity to known killer domains, like those within the Fas/
TNF/FADD/TRADD receptor family (reviewed in [7–12]) or the BH3 domains
of Bax, Bak or Bik [129]. If the C-terminus was a killer domain, residues Y1251
and H1293/K1294 would presumably mitigate the action of this putative killer
domain, since their ablation generates a receptor that is inactive for survival
signaling. A second possibility is that the IGF-IR C-terminus acts as an intrinsic inhibitory domain that suppresses the survival signal, possibly by modulating phosphorylation or dephosphorylation of other receptor residues involved in the survival signal. In this scenario, the critical residues at
Y1250/Y1251 and H1293/K1294 would presumably be involved in neutralizing
the inhibitory action of the C-terminal domain, and C-terminal truncation
mutants lacking this inhibitory domain might be super-active.

Further studies were very recently undertaken in order to investigate the role
of the C-terminus in modulating cell survival [130]. Expression of the IGF-IR
C-terminal 108 amino acids as a membrane targeted protein domain resulted in
induction of apoptosis in MCF-7, thus confirming that it has intrinsic pro-
apoptotic activity. The cytotoxic effect of the C-terminal cytotoxicity could be
abrogated by introducing the Y1250/1251F H1293F/K1294L mutations, (which
abolished suppression of apoptosis in the context of the full-length receptor).
This indicates that the C-terminus may act by exerting a dominant negative
phenotype resulting in sequestration of substrates that are normally required
for endogenous IGF-IR to keep cells alive. In fact expression of the IGF-IR C-terminus in cells results in the same phenotype as expression of antisense oligo
nucleotides, thereby confirming its critical role in regulating the survival activity of the receptor.

6.3
Substrates Interacting with the IGF-IR C-Terminus

The C-terminus of the IGF-IR plays an important role in regulating the survival
activity of the receptor. Therefore it is of great interest to identify what substrates interact with this part of the receptor, in particular with the Y1250/Y1251
and H1293/K1294 regions necessary for survival. The only proteins currently
know to interact with the IGF-IR C-terminus are p85 of PI3-kinase which interacts with Y1316 [108] and grb-10 thought to interact between residues 1229 and
1245 [131]. Substrates that interact with Y1250/1251 or with the H1293/K1294
domains of the receptor have not yet been identified, but based on the requirement for these residues for survival signaling it is likely that they interact with
proteins that are either effectors or regulators of IGF-IR-mediated suppression
of apoptosis.

7
Signal Transduction by Survival Factors

7.1
PI-3 Kinase in Survival

The first indications of which signal transduction molecules might be involved in survival signaling from activated survival factor receptors came from studies of the NGF-Trk receptor system in PC-12 cells [47]. In this system NGF binding to the Trk receptor can both signal cellular differentiation via a ras-dependent pathway and also suppress apoptosis. Using the fungal protein wortmannin and the synthetic molecule LY294002 [132] as specific inhibitors of PI-3 kinase, it was found that suppression of apoptosis by NGF in response to serum withdrawal was dependent on PI-3 kinase activity, whereas neurite out-growth and differentiation was not. In the same system, suppression of apoptosis by PDGF receptors which had been transfected into the cells was also dependent on PI-3 kinase activity. This study firmly implicated PI-3 kinase as a potential mediator of cell survival. Subsequently PI-3 kinase activity was determined to be critical for suppression of apoptosis by survival factors in other cell systems. In Rat-1 fibroblasts wortmannin substantially inhibits protection from c-Myc induced apoptosis by serum and IGF-I [133, 134] and inhibits protection from UV-induced apoptosis by IGF-I [100]. The PI3-kinase inhibitor LY294002 inhibits insulin-mediated survival of cerebellar neurons [135]. Over-expression of a constitutively active PI3-kinase molecule which had a truncated p85 sub-unit linked to the catalytic p110 subunit [136] was sufficient to provide protection from apoptosis induced by UV irradiation in Cos cells [100], and a constitutively activated p110 subunit of PI-3 kinase was sufficient to protect Rat-1 fibroblasts from c-Myc-induced apoptosis [133].

The downstream effector of PI-3 kinase-mediated suppression of apoptosis is apparently the serine threonine kinase AKT or protein kinase B. PI-3 kinase can activate at least two downstream kinases, AKT and the p70 ribosomal S6 kinase. In cerebellar neurons and in Rat-1 fibroblasts, over-expression of AKT could protect from serum withdrawal or Myc-induced apoptosis, respectively [100, 133]. Activation of p70 ribosomal S6 kinase by PI-3 kinase was not associated with cell survival, because inhibition of this enzyme by rapamycin did not affect the protective effects mediated by insulin or serum in neurons or PI3-kinase-mediated suppression of apoptosis by c-Myc in Rat-1 fibroblasts [133].

7.2
AKT and PI-3 Kinase

Protein kinase Bα or RACα or c-AKT named for its sequence homology to protein kinases A and C was isolated in 1991 [137], and was later found to be the cellular homolog of V-Akt, which is a protein encoded by the rodent T cell lymphoma acute transforming retrovirus AKT-8 [138]. Two other isoforms of

PKB called PKB β (RacB, AKT-2) [139] and PKBγ [140] have also been identified. PKB has an N-terminal AKT homology domain (AH) [141], which is homologous to the pleckstrin homology domain present in several signaling proteins [142]. Next to the N-terminal AH domain is a catalytic domain, which has 65 % similarity to protein kinase A and 75 % similarity to protein kinase C, and then a C-terminal tail. AKT becomes phosphorylated within minutes of stimulation by several cytokines including IGF-I, insulin, PDGF, EGF, and basic FGF [143, 144] and this effect is inhibited by PI3-kinase inhibitors; therefore AKT is functionally downstream of PI3-kinase.

The product of PI-3 kinase activity, phospahtidyl inositol-3,4,5 biphosphate (PI-3,4,5-p2) was shown to bind to the pleckstrin homology domain in AKT in vitro and facilitate its dimerization, resulting in autophosphorylation and activation of AKT [145]. AKT may be targeted to the membrane by binding of another PI-3 kinase product PtdIns-3,4,5-P3 [145], a lipid which does not facilitate its dimerization or activation. Versions of AKT that have the viral gag protein gag-AKT or have a sequence for N-terminal myristylation attached are constitutively active [133, 134]. Activation of AKT by insulin or IGF-I results in its phosphorylation on threonine 308 and on serine 473, and these phosphorylations are necessary for it to become fully active [143]. The presence of upstream kinases other than AKT have been proposed to account for this, and presumably these would have to be activated by receptors such as the IGF-IR. Recently one such kinase called phosphoinositide-dependent protein kinase (PDK1) has been purified [146]. PDK1 is activated by phosphoinositides (PtdIns (3,4,5)P3 or PtdIns (3,4)P2 and can in turn activate PKB. It is currently not known how PDK1 might be activated by survival factor receptors.

A downstream substrate of AKT is GSK-3, the phosphorylation of which is inhibited in response to insulin with the same half time of activation and phosphorylation of AKT by insulin [147]. GSK-3 dephosphorylation is associated with stimulation of glycogen synthesis and translation of certain mRNAs by insulin [148]. GSK-3 has also been shown to inhibit translocation of NF-AT into the cell nucleus [149].

AKT-2 (PKBβ, RACβ) has been shown to be over-expressed in a significant number of ovarian and pancreatic cancers [150, 151] and PKBa is amplified in the breast carcinoma cell line MCF-7 [152]. Inhibition of AKT expression by antisense RNA in pancreatic tumor cells lead to inhibition of tumorigenesis [151]. It is not known whether the up-regulation of AKT in tumor cells is dependent on survival factor activity or is independent of survival factors. It would also be interesting to know if and how AKT is regulated during carcinogenic progression. This could be documented in a model system such as the RIP-Tag model of tumorigenesis [111]. With its documented role in suppression of apoptosis and over-expression patterns AKT is an attractive target for therapeutic intervention in tumors. Inhibition of AKT activity should lead to suppression of survival and lead to apoptosis of the tumor cells. However, it remains to be seen whether it will be possible to specifically inhibit this kinase and not the many other similar kinases in the cell.

7.3
AKT as a Mediator of Survival Factor Action

Growth factor receptor activation of PI-3 kinases is thought to be necessary and sufficient for activation of AKT (summarized in [153] and references cited therein). Many survival factor receptors such as the PDGF receptor and the IGF-IR can directly or indirectly recruit and activate PI-3 kinase upon ligand binding. Mutation of the PI-3 kinase tyrosine binding residue in the PDGF receptor has been shown to prevent the activation of AKT [154]. This effect is inhibited by inhibitors of PI-3 kinase such as wortmannin. Therefore, it seems that survival factor receptors activate AKT via direct activation of PI-3 kinase.

IL-2 and IL-3 suppress apoptosis and activate AKT in a PI-3 kinase-dependent manner in the IL-3-dependent cell line BAF/3, an effect that is dependent on the presence of the IL-2 receptor β chain; and by IL-2 in an IL-2-dependent T cell line [155]. In this study, expression of a constitutively active AKT protein resulted in increased levels of Bcl-2 and c-myc expression, and caused cells to progress through the cell cycle. This is likely to be a different signal to that transmitted by AKT through survival factors such as IGF-1, because IGF-I can protect cells from apoptosis in the absence of cell cycle progression (unpublished results).

A dominant negative AKT blocks insulin-dependent survival in neuronal cells [135], suggesting that AKT is essential for the IGF-IR to signal survival. It is possible that receptors such as the IGF-IR have more than one mechanism to regulate or activate AKT either through kinases such as PDK1 or other proteins. Support for this idea comes from the site-directed mutagenesis study with the IGF-IR in suppression of apoptosis [128]. The two IGF-IR mutants Y950F and Y1316F which should have lost their potential to recruit PI 3 kinase through IRS-I or through direct interaction, respectively, did not lose anti-apoptotic function. One explanation for this is that there could be redundancy of substrate interactions. If one of these tyrosines is abolished the other can signal as effectively as if both are present. An alternative explanation is that PI-3 kinase may be activated by the receptor through as yet unidentified mechanisms. In the context of PI-3 kinase/AKT activity the other interesting question that arises from the IGF-IR mutational analysis is: what effect do the mutant receptors that lose suppression of apoptosis (Y1250/1251F and K1293/K1294) have on the downstream signaling molecules? As discussed above, it not known what these regions of the receptor interact with, and it is not immediately obvious how they might affect the activity of PI-3 kinase or AKT. Truncation of the receptor at the C-terminus enhances the anti-apoptotic signaling of the IGF-IR, and therefore it is conceivable that these residues are involved in regulating the kinase activity of the receptor or in regulating the quality or duration of the signals generated by the receptor. They may modulate the duration of AKT activation by the receptor and thereby indirectly impact the survival signal. It is also possible that these domains of the IGF-IR regulate the interaction of other residues of the receptor with PI-3 kinase, AKT, or AKT kinases.

7.4
Suppression of Apoptosis by AKT: Other Pathways

The only know physiological target of AKT at this time is GSK-3 [147, 148]. It is not apparent how activation of this kinase can regulate suppression of apoptosis. Therefore, it is possible that AKT has as yet unidentified targets that are intermediates in regulation or execution of apoptosis. Candidates for such activity could be regulators of the cell death pathway such as the Bcl-2 family [13–17]: either the death suppressers such as Bcl-2 and Bcl-xl, death promoters such as Bax or Bak, or regulators of these such as Bad and Bclxs [13]. These proteins have sites of phosphorylation and relatively little is known about regulation of their activity through phosphorylation. Phosphorylation of Bcl-2 on serine is associated with loss of its anti-death function [157]. Phosphorylation of Bad on serines by IL-3 activation of FL5.12 cells has been shown to sequester it away from Bcl-xl in the mitochondria, by allowing it to interact with a cytosolic protein 14--3-3 [157]. This results in Bcl-xl having more anti-apoptotic activity, and this has been proposed as a potential survival mechanism used by IL-3 in these cells. This is the first report of a direct connection between survival factors and the Bcl-2 family. However, it is not clear which kinase phosphorylates Bad Raf-1 has been shown to phosphorylate Bad in vitro, but on different sites from those phosphorylated in response to IL-3 [158]. This, therefore, makes Raf an unlikely candidate to be the Bad kinase that functions in IL-3-mediated survival signaling.

An interesting observation is that although PKC does not phosphorylate Bad at the correct site for binding 14–3-3 in vitro [157] heart muscle kinase does, suggesting that a PKA-related kinase might be effective in vivo. Of course an obvious candidate to phosphorylate Bad in this manner would be AKT, and, as mentioned above, IL-3 has been shown to activate AKT in BAF/3 cells [155]. It will be intriguing to see in the near future if AKT will be the link between survival factors and the Bcl-2 family. The expression levels of Bcl-xl have been shown to increase dramatically in an Epo-dependent manner during the terminal differentiation of human and mouse erythroblasts [159], and Bcl-xl has been shown to increase during T cell activation [15]. IL-2 can activate PI-3 kinase and AKT [155]. Altogether, these observations suggest that hemopoietic cell survival factors (IL-3, Epo, IL-2) have the ability both to modulate Bcl-2 family expression and to activate AKT, and it remains to be seen whether these events are linked, or whether there are redundant pathways.

Survival factors such as IGF-I can block apoptosis induced by activation of the caspase cysteine protease, and therefore it is also possible that AKT could interact with pathways that feed directly into the caspase cascade, for example the pathway initiated at the Fas/TNF receptors [160]. Another intriguing possibility is that AKT could regulate suppression of apoptosis through a functional interaction with the recently cloned ced-4 mammalian homolog Apaf-1 [161]. Ced-4 forms a complex with ced-3 and ced-9 and in the worm is thought to be responsible for activation of ced-3 and triggering of apoptosis [162]. In mammalian systems Apaf-1 is thought to have a similar function by activating caspase 3 [161], and like other proteins involved in regulating apoptosis it is likely to be the

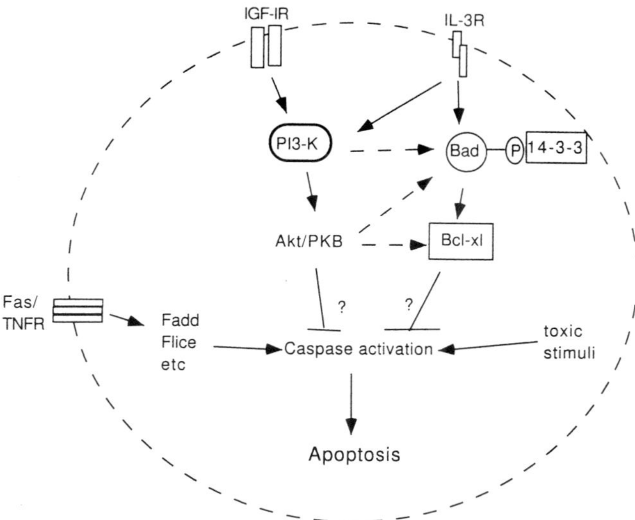

Fig. 7. Schematic representation of survival signaling in cells. Signals from the IGF-IR and IL-3R can protect cells from multiple pro-apoptotic signals that lead to activation of caspases and apoptosis. Signals from the IGF-IR activate PI-3 kinase and AKT to promote survival IL-3 can activate PI-3 kinase and AKT, but also phosphorylates the Bcl-2 family member Bad, resulting in its sequestration by 14-3-3 and the liberation of Bcl-xl to promote survival. The ultimate result of both of these pathways presumably leads to blockage of caspase activation. However, how AKT or Bcl-xl (and Bcl-2 family members) block caspase activation is not currently know. It is also not known whether there is interaction between the PI-3 kinase/AKT pathway and Bcl-2 family members. Putative functional connections between these pathways are indicated by *dashed arrows* (see note in proof)

first of a family of ced-4 homologs. A schematic representation of survival signaling pathways in shown in Fig. 7.

8
Summary and Future Perspectives

Suppression of apoptosis by survival factors is critical for normal development, for the survival of neurons and other normal cells, and also contributes to the evolution and maintenance of tumors. It appears that the IGF-I/IGF-IR system provides the most widely expressed survival signals in the body, and at least in some systems mediates survival through activation of PI-3 kinase and AKT. However, IGF-I and other survival factors can also modulate expression of Bcl-2 family members, and it remains to be seen if and how the PI-3 kinase signaling is connected to the regulators and effectors of apoptosis such as the Bcl-2 family, a putative Apaf-1 family, caspases, and the TNFR/Fas pathways.

Activation of IGF-IR protects damaged neurons from cell death, and inhibition of IGF-IR function induces apoptosis in tumor cells. Therefore, there are possibilities for therapeutic intervention by modulating cell death in diseases where there is aberrant cell survival, such as death of neurons due to ischemia, or the development and persistence of tumors due to increased expression of IGF-IR. The challenge remains to target specifically the diseased cells and not the surrounding normal cells. Prospects for a therapeutic window are enhanced by the fact that tumor cells or damaged neurons seem to be more dependent on receiving specific survival signals than the surrounding normal tissues, because they up-regulate the IGF-IR, its ligands, or other proteins involved in survival signaling.

Acknowledgements. The author wishes to thank Yimao Liu and my colleagues at Apoptosis Technology Inc, Cambridge, MA for their contribution to the work cited, and for many fruitful scientific discussions.

Note Added in proof: It has now been shown by at least 2 groups that AKT can phosphorylate Bad on serine (see refs 163, 164).

9
References

1. Kerr JFR, Wyllie AH, Currie AR (1972) Br J Cancer 26:1749
2. Duvall E, Wyllie AH (1986) Immunol Today 7:115
3. Alnemri ES, Livingston DJ, Nicholoson DW, Salvesen G, Thornbery NA, Wong WW, Juan J (1996) Cell 87:171
4. Yuan J, Shaham S, Ledoux S, Ellis HM, Horovitz HR (1993) Cell 75:641
5. Thornberry NA, Bull HG, Calaycay JR, Chapman KT, Howard AD, Kostura MJ, Miller DK, Molineaux SM, Weidner JR, Aunins J, Elliston KO, Ayala JM, Casano FJ, Chin J, Ding G, Egger LA, Gaffney EP, Limjuco G, Palyha OC, Raju SM, Rolando AM, Salley JP, Yamin TT, Lee TD, Shively JE, MacCross M, Mumford RA, Schmidt JA, Tocci MJ (1992) Nature 356:768
6. Tewari M, Quan LT, O'Rourke K, Desnoyers S, Zeng Z, Beidler DR, Poierier GG, Salvesen GS Dixit VM (1995) Cell 81:801
7. Itoh N, Yonehara S, Ishii A, Yonehara M, Mizushima S-I, Sameshima M, Hase A, Seto Y, Nagata S (1991) Cell 66:233
8. Oehm A, Behrmann I, Falk W, Pawlita W, Maier G, Klas C, Li-Weber M, Richards S, Dhein J, Trauth BC, Ponstigl H, Krammer PH (1992) J Biol Chem 267:10,709
9. Suda T, Takahashi T, Golstein P, Nagata S (1993) Cell 75:1169
10. Tartaglia LA, Ayres TM, Wong GHW, Goeddel DV (1993) Cell 74:845
11. Pan G, Ni J, Wei YF, Yu G-L Gentz R, Dixit VM (1997) 277:815
12. Sheridan JP, Marsters SA, Pitti RM, Gurney A, Skubatch M, Baldwin D, Ramakrishnan L, Gray CL, Baker K, Wood WI, Goddard AD, Godowski P, Ashkenazi A (1997) Science 277:818
13. Yang E, Korsmeyer SJ (1996) Blood 88:386
14. Vaux DL, Cory S, Adams JM (1988) Nature 335:440
15. Boise LH, Gonzales-Garcia M, Postema CE, Ding L, lindsten T, Turka LA, Mao X, Nunez G (1993) Cell 74:597
16. Oltvai ZN, Milliman CN, Korsmeyer SJ (1993) Cell; 74:609
17. Chittenden T, Harrington EA, O'Connor R, Flemington C, Lutz RJ, Evan GI, Guild B (1995) Nature 374:733
18. Harrington E, Bennett M, Fanidi A, Evan GI (1994) EMBO J, 13:3286–3295
19. Ellis HM, Horvitz HR (1986) Cell 44:817
20. Horovitz HR, Shaham S, Hengartner MO (1994) Cold Spring Harbour Symp Quant Biol 59:377
21. Hengartner MO, Horvitz, HR (1994) Cell 76:665

22. Shaham S, Horovitz HR (1996) Cell 86:201
23. Vaux DL, Weissman IL, Kim SK (1992) Science 258:1955
24. Ffrench-Constant C (1992) Current Biol 2:577
25. Hockenberry D, Zutter M, Hickey W, Nahm M, Korsmeyer SJ (1991) Proc Natl Acad Sci 88:6961
26. Strasser A, Harris AW, Cory S (1991) Cell 67:889
27. Liu JP, Baker J, Perkins AS, Robertson EJ, Efstratiadis A (1993) Cell 75:59
28. Raff MC (1992) Nature 356:397
29. Williams GT, Smith CA, Spooncer E, Dexter TM, Taylor TR (1990) Nature 343:76
30. Fairbairn LJ, Cowling GJ, Reipert BM, Dexter TM (1993) Cell 74:823
31. Wu H, Liu X, Jaenisch R, Lodish HF (1995) Cell 83:59
32. Kourry X, Bondurant X (1990) Science 248:278
33. Hamburger V, Levi-Montalcini R (1949) J Exp Zool 111:457
34. Glucksman A (1951) Biol Rev Camb Philos Soc 26:59
35. Saunders JW (1966) 154:604
36. Ockel M, von-Schack D, Schropel A, Dechant G, Lewin GR, Barde YA (1996) Phiolos-TransR-Soc-Lond-B-Biol-Sci 351:383–387
37. Lindsay RM (1996) Phiolos-Trans-R-Soc-Lond-B-Biol-Sci 351:365–373
38. Lewin GR (1996) Neurons and the specification of neuronal phenotype. Phiolos-Trans-R-Soc-Lond-B-Biol-Sci 351:405–411
39. Levi-Montalcini R (1987) Science 237:1154
40. Glass DJ, Yancopoulos GD (1993) Trends Cell Biol 3:292
41. Kaplan DR, Martin-Zanca D, Parada LF (1991) Nature 350:158
42. Rodriguez-Tebar A, Dechant G, Barde YA (1990) Neuron 4:487
43. Chao MV, Hempstead BL (1995) Trends in Neurosci 18:321
44. Barbacid M (1994) J Neurobiol 25:1386–1403
45. Bothwell M (1991) Current Topic Microbiol Immunol 165:55–70
46. Thomson TM, Chao MV, Hempstead BL (1995) Trends in Neurosci 18:321–326
47. Yao R, Cooper JM (1995) Science 267:2003
48. Couin A, Camu W, Bloch-Gallego E, Mettling C, Henderson CE (1993) C-R-Seances-Soc-Biol-Fil 187:47–61
49. Hughes RA, Sendter M, Thoenen H (1993) J Neurosci Res 15:663–671
50. Koliatsos VE, Cayouette MH, Berkemeier LR, Clatterbuck RE, Price DL, Rosenthal A (1994) Neurotrophin4/5 is a trophic factor for mammalian facial motor neurons. Proc Nat Acad Sci USA 91:M 3304–3308
51. D'Mello SR, Gali C, Ciotti T, Calissano P (1993) Proc Natl Acad Sci USA 90:10,989–10,993
52. Poulsen KT, Armanini MP, Klein RD, Hynes MA, Phillips HS, Rosenthal A (1994) Neuron 13:1245–1252
53. Raff Mc, Barres BA, Burne JF, Cole HS, Ishikazi Y, Jacobson MD (1993) Science 262:695
54. Greene LA (1978) J Cell Biol 78:747
55. Ishizaki Y, Voyvodic JT, Burne JF, Raff MC (1993) J Cell Biol 121:899
56. Masuda K, Watanabe I, Unoki K, Ohba N, Muramatsu T (1995) Invest Opthamol Vis Sci 36:2142
57. Steinberg, RH (1994) Curr Opin Neurobiol 4:515
58. Hseuh AJ, Eisenhauer K, Chun SY, Hsu SY, Billig H (1996) Recent Prog Hormon Res 51:433
59. Chun SY, Eisenhauer KM, Minami S, Billig H, Perlas E, Hseuh AJ (1996) Endocrinology 137:1447
60. Isaacs JT, Furuya Y, Berges R (1994) Semin Cancer Biol 5:391
61. Metcalf D (1985) Blood 229:16
62. Mekori YA, Oh CK, Metcalfe DD (1993) J Immunol 151:3775–3784
63. Caceres-Cortez J, Rajotte D, Dumouchel J, Haddad P, Hoang T (1994) J Biol Chem 269:12,084–12,091
64. Collins MK, Marvel J, Malde P, Lopez-Rivas A (1992) J Exp Med 176:1043–1051
65. Collins MK, Perlins GR, Rodriguez-Tarduchy R, Nieto MA, Lopez-Rivas A (1994) Bioessays 16:133–138

66. Wu H, Liu X, Jaenisch R, Lodish HF (1995) Cell 83:59
67. Socolvsky M, Dusanter-Furt I, Lodish H (1997) J Biol Chem 272:14,009–14,012
68. Zubiago AM, Munoz E, Huber BT (1992) J Immunol 149:107–112
69. Rinderknecht E, Hummel RE (1976) Proc Natl Acad |Sci USA 73:4379
70. Hummel RE (1989) Eur J Biochem 89:1540
71. Foyt H, Roberts CT (1991) In: LeRoith (ed) Insulin-like growth factors. CRC Press, Boca Raton, Fl, pp1–16
72. Tricoli JV, Rall LB, Scott J, Bell GI, Shows TB (1984) Nature 310:784
73. Adamo ML Ben-Hur H, LeRoith D, Roberts CT (1991) Mol Endocrinoo 5:1677
74. Salmon WD Jr, Daughaday WH (1957) J Lab Clin Med 49:825
75. Melmed S, Yamashita S, Yamasaki H, Fagin J, Namba H, Yamamoto H, Weber M, Morita S, Webster J, Prager (1996) Recent Prog Horm Res51:189
76. Backstrom M, Hall K, Sara VR (1984) Acta Endocrinol 197:171
77. Baxter RC, Martin JI (1989) Prog Growth Factor Res 1:49
78. Rajah R, Valentis B, Cohen P (1997) 272:12,181
79. Morgan DO, Edman JC, Standring DN, Fried V, Smith MC, Roth RA, Rutter WJ (1987) Nature 329:301
80. Baserga, R, Sell, C, Porcu, P, Rubini M (1994) Cell Prolif 27, 63
81. Pietrzkowski Z, Lammers R, Carpenter G, Soderquist AM, Limardo M, Philips PD, Ullrich A, Baserga R (1992) Cell Growth and Diff 3:199
82. Steller MA, Delgado CH, Zou Z (1995) Proc Natl Acad Sci 92:11,970
83. Rodriguez-Tarduchy G, Collins MKL, Garcia I, Lopez-Rivas A (1992) J Immunol 149:535
84. O'Connor R, Cesano A, Lange B, Finan J, Nowell PC, Clark SC, Raimondi SC, Rovera G, Santoli D (1991) Blood 77:1534
85. O'Connor R, Torigoe T, Reed JC, Santoli D (1992) Blood 80:1017
86. Goldring B, Goldring, SR (1991) Crit Rev Eukaryotic Gene Expression 1:301
87. Evan GI, Wyllie AH, Gilbert CS, Littlewood TD, Land H, Brooks M, Waters CM, Penn LJZ, Hancock DC (1992) Cell, 69:119
88. Fanidi A, Harrington E, Evan G (1992) Nature, 359, 554–556
89. Barres BA, Schmid R, Sendnter M, Raff MC (1993) Development 118:283–295
90. Mathews CC, Feldman EL (1996) J Cell Physiol 166:323
91. Gluckman P, Klempt N, Guan J, Mallard C, Sirimanne E, Dragunow M, Klempt M, Singh K, Williams C, & Nikolics K (1992) Biochem Biophys Res Commun 182:593
92. Cheng B, Mattson MP (1992) J Neurochem 12:1558
93. Gluckman PD, Guan J, Beilharz EJ, Klempt ND, Klempt M, Miller O, Sirimanne E, Dragunow M, Williams CE (1993) Annals NY Acad Sci 692:138
94. Research News (1994) Science 264:772
95. LeRoith D, Werner H, Beitner-Johnson D, Roberts CT (1995) Endocrine Revs 16:143
96. Werner H, LeRoith D (1996) Adv Cancer Res 168:183
97. Christofori G, Naik P, Hanahan D (1994) Nature, 369, 414
98. Zhang L, Zhou W, Velculescu VE, Kern SE, Hruban RH, Hamilton SR, Vogelstein B, Kinzler KW (1997) Science, 276:1268
99. Sell C, Baserga R, Rubin R (1995) Cancer Res 55:303
100. Kulik G, KlippelA, Weber MJ (1997) Mol Cell Biol 17:1595–1605
101. Jung Y-K, Miura M, Yuan J (1996) J Biol Chem 271:5112–5117
102. Abbot AM, Bueno R, Pedrini MT, Murray JM, Smith RJ (1992) J Biol Chem 267:10,759
103. Werner H, Karnoelli E, Rauscher III FJ, LeRoith D (1996) Proc Natl Acad Sci 93:8318
104. Werner H, Karnieli E, Rauscher II, FJ, LeRoith D (1996) Proc Natl Acad Sci 93:38,318
105. Bondy CA, Werner H, Roberts CT Jr, LeRoith D (1989) Mol Endocrinol 4:1386
106. Werner H, Woloschak M, Adamo M, Shen-Orr Z, Roberts CT Jr, LeRoith D (1989) Proc Natl Acad Sci USA 86:7451
107. Ullrich A, Gray AW, Yang-Feng M, Tsubokawa C, Collins W,Henzel T, LeBon S, Katchuria S, Chen E, Jakobs S, Francke Y, Ramachandran J, Fujita-Yamaguchi Y (1986) EMBO J, 5:2503

108. Gustafson TA, He W, Crapara A, Schaub CD, O'Neill TJ (1995) Mol Cell Biol 15:2500
109. Yamamoto K, Altschuler, D, Wood E, Horlick K, Jacobs S, Lapetino EG (1992) J Biol Chem 267:11,337
110. Hanahan D (1985) Nature 315:115
111. Naik P, Karrim J, Hanahan D (1996) Genes and Dev 10:2105
112. Strasser A, Harris AW, Bath ML, Cory S (1990) Nature 348:331
113. Adams JM, Harris AW, Pinkert CA, Corcora LM, Alexander WS, Cory S, Palmitter RD, Brinster RL (1985) Nature 318:533
114. Kaleko M, Rutter WJ, Miller AD (1990) Mol Cell Biol, 10:464
115 Peterson JE, Jelinek T, Kaleko M, Siddle K, Weber MJ (1994) J Biol Chem 269:27,315
116. Sell C, Dumenzil G, Deveaud C, Miura M, Coppola D, DeAngelis T,. Rubin R, Efstratiadis A, Baserga R (1994) Mol Cell Biol 14:3604
117. Surmacz E, Sell C, Swantek J, Kato H, Roberts CTJ, LeRoith D, Baserga R (1995) 218:370
118. Arteaga CL (1982) Breast Cancer Res Treat 22:101
119. Pietrzkowski Z, Mulholland G, Gomella L, Jameson BA, Wernicke D, Baserga R (1993) Cancer Res 53:1102
120. Resnicoff M, Abraham D, Yutanawiboonchai W, Rotman H, Kajstura J, Rubin R, Zoltick P, Baserga R (1995) Cancer Res, 55:2463
121. Rescinoff M, Burgaud J, Rotman HL, Abraham D, Baserga R. (1995) Cancer Res 55:3739
122. Burfeind P, Chernicky CL, Rinisland F, Ilan J, Ilan J (1996) Proc Natl Acad Sci 93:7263
123. Lee CT, Wu S, Gabrilovich D, Chen H, Dadaf-Rahrov S, Ciernik IF, Carbone DP (1996) Cancer Res 56:3038
124. Prager D, Li HL, Asa S, Mehmed S (1994) Proc Natl Acad Sci 91:2181
125. Burgaud J-L, Resnicoff M, Baserga R (1995) Biochem Biophys Res Commun 214:475
126. Hongo A, D'Ambrosio C, Miura M, Morrione A, Baserga R (1996) Oncogene, 12:1231
127. Li S, Ferber A, Miura M, Baserga R (1994) J Biol Chem 271:32,558
128. O'Connor R, Kauffmann-Zeh A, Liu Y, Lehar S, Eva GI, Baserga R, Blattler WA (1997) Mol Cell Biol 17:427
129. Chittenden T, Flemington C, Houghton AB, Ebb RG, Gallo GJ, Elangovan B, Chinnadurai G, Lutz RJ (1995) EMBO J 14:5589
130. Liu Y, Lehar S, Corvi C, Payne G, O'Connor R (1998) Expression of the IGF-I Receptor C-terminus as a myristylated protein leads to induction of apoptosis in tumor cells Cancer Res 58:570–576
131. Morrone A, Valentis B, Li S, Ooi JYT, Margolis B, Baserga R (1996) Cancer Res 56:1
132. Vlahos CJ, Matter WF, Hui HK, Brown RF (1994) J Biol Chem 269:5241
133. Kauffmann-Zeh A, Rodriguez-Viciana P, Ulrich E, Gilbert C, Coffer P, Downward J, Evan G (1997) Nature 385:544
134. Kennedy SG, Wagner AJ, Conzen SD, Jordan J, Bellacosa A, Tsichlis PN, Hay N (1997) Genes and Dev 11:701
135. Dudek H, Datta SR, Franke TF, Birnbaum MJ, Yao R, Cooper GM, Segal R, Kaplan DR, Greenberg ME (1997) Regulation of neuronal survival by the serine threonine kinase. AKT Science 275:661
136. Hu Q, Klippel A, Muslin AJ, Fantl WJ, Williams LT (1995) Science 268:100
137. Coffer P, Woodgett JR (1991) Eur J Biochem 201:475
138. Staal SP, Hartley JW, Rowe WP (1977) Proc Natl Acad Sci USA 74:3065
139. Jones PF, Jakubowicz T, Pitossi FJ, Maurer F, Hemmings B (1991) Proc Natl Acad Sci USA 88:4171
140. Konishi H, Kuroda S, Tanaka M, Matsuzaki H, Ono Y, Kameyama K, Haga T, Kikkawa U (1995) Biochem Biophys Res Commun 216:526
141. Bellacosa A, Franke TF, Gonzales-Portal ME, Datta K, Taguchi T, Gardner J, Cheng JQ, Testa JR, Tsichlis PN (1993) Oncogene 8:745
142. Cohen GB, Ren R, Baltimore D Cell (1995) 80:237
143. Alessi DR, Mirjana A, Caudwell B, Cron P, Morrice N, Cohen P, Hemmings BA (1996) EMBO J 15:6541

144. Franke TF, Yang S, Chan TO, Datta K, Kazlauskas A, Morrison DK, Kaplan DR, Tsichlis PN (1995) Cell 81:727
145. Franke TF, Kaplan DR, Cantley LC, Toker A (1997) Science 275:665
146. Alesi DR, James SR, Downes CP, Holmes AB, Gaffney PR, Reese CB, Cohen P (1997) Curr Biol 7:261
147. Hemmings BA (1997) Science 275:628
148. Cross DA, Watt PW, Shaw M, van der Kaay J, Downes CP, Holder JC, Cohen P (1997) Febbs Lett 7:211
149. Beals CR, Sheridan CM, Turck CW, Gardner P, Crabtree GR (1997) Science 275:1930
150. Cheng JQ, Godwin AK, Bellacosa A, Taguchi T, Franke TF, Hamilton TC, Tsichlis PN, Testa JR (1992) Proc Natl Acad Sci Usa 89:9267
151. Cheng JQ, Ruggeri B, Klein WM, Sonoda G, Altomare DA, Watson DK, Testa JR (1996) Proc Natl Acad Sci USA 93:3636
152. Jones PF, Jakubowicz T, Hemmings BA (1991) Cell Regul 2:1001
153. Franke TF, Kaplan DR, Cantley LC (1997 Cell 88:433
154. Burgering BMT, Coffer PJ (1995) Nature 376:599
155. Naheed N, Ahmed L, Grimes L, Bellacosa A, Chan TO, Tsichlis PN (1997) Proc Natl Acad Sci 94:3627
156. Haldar S, Jena N, Croce CM (1995) Proc Natl Acad Science USA, 92:4507
157. Zha J, Harada H, Yang E, Jockel J, Korsmeyer S (1996) Cell 87:619
158. Wang HG, Rapp UR, Reed JC (1996) Cell 87:629
159. Gregoli PA, Bondurant MC (1997) Blood 90:630
160. Nagata S (1997) Cell 88:355
161. Zou H, Henzel WJ, Liu X, Lutschg A, Wang X (1997) Cell 90:405
162. Wu D, Wallen HD, Inohara N, Nunez G (1997) J Biol Chem 272:21,449
163. Datta SR, Dudek H, Tao X, Masters S, Fu H, Gotoh Y, Greenberg ME (1997) Cell 91:231
164. del Peso L, Gonzalez-Garcia M, Page C, Herrera R, Nunez G (1997) Science 278:687

Received November 1997

Apoptosis and Bioprocess Technology

R.P. Singh · M. Al-Rubeai

Centre for Bioprocess Engineering, School of Chemical Engineering, University of Birmingham, Edgbaston, Birmingham, B15 2TT, UK, *E-mail: m.al-rubeai@bham.ac.uk*

Optimisation of the production of biopharmaceuticals in animal cell lines has become a key area of research. The identification of apoptosis as the major mechanism of cell death during such processes has raised the importance of studies of cell death when implementing culture optimisation strategies. In this article we present an overview of the studies which have demonstrated the induction of apoptosis during the cultivation of industrially important animal cell lines. We also discuss studies which have shown that deprivation of factors such as amino acids, glucose, serum and oxygen are potent inducers of apoptosis in industrial cultures. The suppression of apoptosis under these conditions has been demonstrated by a number of recent reports, and we describe ways in which this knowledge may be applied in the developement of novel solutions to some of the technical problems associated with the development of successful large scale culture process. The article concludes with a discussion of future directions for apoptosis research in bioprocess technology.

Keywords: Animal cell culture, Apoptosis, process optimisation, bcl-2, Tissue engineering, Bioreactor, Culture Media.

Advances in Biochemical Engineering/
Biotechnology, Vol. 62
Managing Editor: Th. Scheper

1
Introduction

Animal cell lines have been increasingly utilised as hosts for the production of recombinant proteins. Primarily, this reflects the need to produce complex proteins for use as therapeutics to a very high degree of fidelity, and has been facilitated by the rapid progress in molecular biology and biomedical research. Consequently, animal cell culture has become a multi-billion dollar sector of biotechnology. With many potential products in the clinical trials pipeline, it is predicted that, by the year 2003, recombinant proteins will constitute approximately 10 % of the projected world pharmaceuticals market [1]. As a result, considerable research effort has been directed towards the development and optimisation of the cell culture process with the objective of delivering the quantity and quality of protein required by the healthcare industries.

Central to the optimisation process is the study of the interactions between cells and the chemical, biochemical and physical components of the culture environment, and the impact of these factors on protein productivity, cell proliferation and cell death. Previously, as with most areas of biological research, the terms cell death and necrosis were used interchangeably by cell culture scientists. However, it is now clear that many of the mammalian cell lines which are used for industrial scale recombinant protein production actually undergo apoptotic death in the bioreactor environment. The high susceptibility to apoptosis may provide a partial explanation for many of the technical problems that are associated with large scale animal cell cultivation.

In this article we will begin by describing the basic morphology of apoptosis and the techniques that are used to identify the mechanism of cell death in the in-vitro culture environment. We will then describe the ground-breaking studies which have investigated the extent of apoptosis during the cell culture process. This will be followed by a discussion of the application of fundamental apoptosis research to the optimisation of the culture process. In this respect, we will draw upon much of the work presented previously. We will conclude with a prediction of future applications of apoptosis research within animal cell technology and in the wider biotechnology sector.

2
Apoptosis: Morphology and Identification in Industrially Important Mammalian Cell Lines

2.1
Morphology

The following description is based on studies of hybridoma cells. It should be noted that there may be variations in the detailed morphology from one cell line to the next. The early stages of apoptosis are characterised by a reduction in cell volume [2, 3] and loss of surface microvilli. Large protrusions form on the outer surface in a process referred to as blebbing. These protrusions may break-away as intact vesicular structures called "apoptotic bodies" [4]. The resultant structures may retain plasma membrane integrity for a number of hours and usually contain chromatin fragments and/or organelles.

During apoptosis, the large nucleus which is seen in many viable cells condenses and marginates to the nuclear membrane forming crescent or ring-shaped structures. This appears to be a relatively short lived morphological phase, with the chromatin eventually forming two or more dense spherical masses (for example see [5]). The shape of the cell also undergoes a highly characteristic change. Whilst viable cells tend to be irregular in shape, on entry into apoptosis they become smooth-surfaced and in many cases perfectly spherical.

The changes in nuclear morphology coincide with endonuclease activation which cleaves chromatin firstly into 300 and/or 50 Kbp fragments, and then ultimately into multiples of 200 bp [6, 7]. The latter of the two stages represents cleavage at the internucleosomal sites and is responsible for the striking ladder-like pattern when DNA samples from apoptotic cells are subjected to DNA gel electrophoresis, which has become a key assay for apoptotic cells.

Retention of the structural integrity of apoptotic cells facilitates their rapid and efficient removal by phagocytic cells, which is vitally important in tissues undergoing high levels of apoptosis. This ensures that the toxic by-products of cellular breakdown are not released into the surroundings, which would result in high levels of damage to neighbouring tissue. This stabilisation of the dying cell is thought to result from the activation of transglutaminase [8], an enzyme which cross-links proteins within the cell generating a protein scaffold which is believed to hold the dead cell together.

2.2
Identification

The two primary techniques used for the identification of apoptosis are DNA gel electrophoresis as described above, or visual confirmation of apoptotic nuclei morphology. Quantification of apoptosis is best carried out by fluorescence microscopic examination of nuclear morphology of cells following staining with a DNA fluorochrome such as acridine orange [9]. Cells can be fixed and stored in formaldehyde at 4°C for extended periods prior to analysis. Alternatively, immediate analysis of cells whilst still in culture medium by simultaneous use of acridine orange and propidium iodide provides additional information regarding plasma membrane integrity [10]. This technique makes use of the differential membrane permeability of these two stains. Acridine orange is permeable to the intact plasma membrane of viable and early apoptotic cells, and consequently these will fluoresce green. Following damage to the plasma membrane, late apoptotic and necrotic cells become permeable to propidium iodide and therefore exhibit red fluorescence.

The possibility of obtaining detailed information on the morphological and biochemical changes associated with apoptosis at the level of the individual cell has stimulated tremendous interest in the development of flow cytometric assays. At present, there are a number of techniques that are available, and some of the most commonly used are discussed below. Although these techniques are widely used, it is inadvisable to rely on one technique in isolation. In many cases, necrotic and late apopotic cells will give similar results, and it is therefore important that results are confirmed by one (or more) other technique(s), such as DNA gel electrophoresis or fluorescence microscopy.

2.2.1
Light Scattering Properties

Changes in cell morphology and size can be revealed by flow cytometric analysis of changes in light scattering properties. The increased granularity caused by nuclear condensation produces an increase in orthogonal light scatter [11, 12]. The reduction in cell size is reflected in a reduction in forward scattered light. The technique can also be used for the identification of necrotic cells (at least in their early stages). When necrosis is induced by a permeablising agent such as saponin, a reduction in forward scatter is seen, but the increase in side scatter which occurs during apoptosis is not observed.

2.2.2
Changes in DNA Content

Leakage of cleaved DNA from apoptotic cells undergoing secondary necrosis results in a reduction in the cellular DNA content. This can be revealed by flow cytometric analysis of cells stained with a DNA fluorochrome such as propidium iodide [13–15]. Apoptotic cells appear as a distinct population below the G1 phase of viable cells (often referred to as the sub-G1 peak). The technique provides a rather good correlation with the fluorescence microscopic method described earlier. However, there are two potential drawbacks to this technique. First, necrotic cells can also exhibit a reduction in DNA content and therefore a sub-G1 peak. Second, the technique is most effective under conditions in which large numbers of cells enter apoptosis in a relatively short period, as this generates a well-defined sub-G1 peak which can be easily quantified.

2.2.3
Annexin V

Phosphatidylserine (PS) is located on the inner leaflet of the plasma membrane in viable cells. Loss of this assymetrical distribution is one of the earliest changes which take place during the onset of apoptosis. [16, 17]. These changes can be revealed by a highly effective flow cytometric assay which makes use of annexin V, a naturally occurring chemical which has high affinity for PS. This is achieved by conjugation of annexin V to FITC, which fluoresces green when excited with laser light at 488 nm. By combining this method with PI staining, it is possible to classify apoptotic cells into two sub-populations: early (membrane intact and therefore PI negative) and late (membrane damaged and therefore PI-positive). The technique has been found to give good correlation with levels of apoptosis during hybridoma batch cultures, during which apoptosis accounts for around 90 % of cell deaths, as revealed by the fluorescence microscopic method described above. However, necrotic cells can also give a false positive result. This is because damage to the plasma membrane allows the annexin to enter the dead cell and bind to PS residues on the inner surface of the plasma membrane. Thus, the technique is only completely reliable as an identification tool when used to analyse early apoptotic cells.

3
Apoptosis and Large Scale Animal Cell Culture

The technical challenges associated with large scale animal cell culture can be classified into 6 groups:

a. the need for maintenance of an optimal culture environment;
b. susceptibility of cells to hydrodynamic stress;
c. anchorage dependence of many of the cell types;
d. the requirement for complex culture medium;
e. low cell densities;
f. low specific productivity.

The development of successful large scale animal cell cultures has been dependent upon the identification of solutions to the above problems. In the discussion that follows, the role of susceptibility to apoptosis in each area will be outlined. The implications of the control of apoptosis will then be demonstrated with reference to specific studies.

3.1
The Need for Maintenance of an Optimal Culture Environment

During the cultivation of animal cell lines, any deviation from an optimal culture environment quickly results in the loss of viability. This characteristic appears to be most obvious in cell lines which exhibit a high sensitivity to apoptosis. In order to assess the level of sensitivity, studies have focused on cell death following sub-optimal conditions: cell death during the decline phase of batch culture and cell death following deprivation of key factors such as amino acids, glucose, oxygen and serum. These studies are discussed further below.

3.1.1
Cell Death During Batch Culture

The growth profile of a batch culture can be divided into four distinct phases. There may be an initial lag phase, during which the cells adapt to the new culture conditions following inoculation. The duration of this phase is highly variable and is dependent upon the immediate culture history of the cells. This is followed by the exponential phase during which cell numbers rapidly increase. The final cell number and the growth rate are dependent upon the cell type, the culture medium used and the general culture conditions. Once maximum cell number has been attained, the culture may enter a stationary phase during which the viable cell number remains fairly steady. The duration of this phase is highly variable and, at least in part, reflects the general robustness of the cell line. Indeed, in many cell lines, such as hybridomas, a stationary phase will not be seen and cultures will rapidly enter the decline phase. The trigger of the decline phase is usually the exhaustion of key nutrients such as glutamine (which is required as an energy source and as a precursor for nucleotide synthesis, in addition to its obvious role in protein biosynthesis) and glucose (which is also required as an energy source). The accumulation of toxic metabolites such as ammonia and lactic acid also leads to the induction of cell death.

Fluorescence and electron microscopic examination of nuclear morphology and DNA gel electrophoresis has revealed high levels of apoptosis during the death phase of batch cultures of murine hybridoma and plasmacytoma (or myeloma) cell lines [9, 18–20]. Studies of *Spodoptera frugiperda* (Sf9) insect cells have revealed predominantly necrotic cell death [9, 21]. During CHO K1 cultures, necrosis has been reported to predominate [9]. Studies of CHO DUX cells in protein-free batch cultures showed clear apoptotic morphology and gel electrophoresis revealed nuclease mediated DNA cleavage [22]. However, apoptotic morphology and DNA cleavage was absent from a CHO cell line, although it was reported that the use of a TUNEL assay gave positive results [23]. It would appear, therefore, that significant variation in the level, morphology and biochemical characteristics of apoptosis exists between different CHO cell lines. Furthermore, studies in our laboratory have revealed that the induction of apoptosis by agents such as staurosporin produces extremely clear characteristic features of apoptosis, which are not readily observed in response to factors which are important in the bioreactor environment. This may be because such factors are not potent inducers of apoptosis in all CHO cell lines.

With the demonstration that the stress factors during the decline phase of batch culture can induce high levels of apoptosis, research effort has focused on strategies which may limit the induction of apoptosis under these conditions. Such strategies have involved genetic suppression of the apoptotic pathway by expression of inhibitor genes, such as those listed in Table 1, or the inclusion of apoptosis suppressor chemicals, such as those listed in Table 2. For example, the overexpression of bcl-2 in hybridoma cultures results in the suppression of

Table 1. Some of the genes involved in apoptosis

Inducer	References	Suppressor	References	Mediator	References
Bcl-Xs	58	Bcl-2	60	Endonucleases	63–65
Fas Receptor/ Ligand	59	Bcl-XL	58	Caspases	66
		p35	61	Trans-glutaminases	67
		BHFR1	62		

Table 2. Selected chemicals which have been demonstrated to inhibit apoptosis induced during in-vitro cultivation of mammalian cells

Inhibitor	References
N-Acetyl-l-cysteine	68
L-Ascorbic acid	32
Aurintricarboxylic Acid	69, 70
Caffeine	71
Calpain Inhibitor I (N-Ac-Leu-Leu-norleucinal)	72, 73
Calpain Inhibitor II (N-Ac-Leu-Leu-normethioninal)	74
Catalase	32
ICE Inhibitor I (Ac-Tyr-Val-Ala-Asp-aldehyde)	75
ICE Inhibitor II (Ac-Tyr-Val-Ala-Asp-CMK)	76
ICE Inhibitor III (Ac-Tyr-Val-Ala-Asp-accyloxymethylketone)	77
Superoxide Dismutase	32

apoptosis during the death phase of batch cultures, leading to a significant improvement in antibody productivity [10, 24]. When bcl-2 transfected COS-1 cells were transfected with the vector pcDNA-lambda carrying the immunoglobulin lambda gene for transient expression of lambda protein, protein expression was higher than that observed in the bcl-2 negative transfected cells. In the same study, the mouse myeloma p3-X63-Ag.8.653, which is used as a fusion partner in the generation of hybridomas, was transfected with the human bcl-2 gene, which resulted in a significant reduction in the death rate. A further enhancement in survival and antibody production in hybridoma 2E3 cultures was observed when cells were co-transfected with bcl-2 and bag-1 [25]. The anti-apoptosis gene E1B-19Kd from Epstein Barr Virus has been transfected into an NS0 cell line which expresses a monoclonal antibody [26]. Although culture duration was extended by many days, antibody productivity was actually lower than in control cultures.

In contrast to these promising studies, transfection of NS0 cells with bcl-2 was reported to have failed to provide any protective effect. Although no endogenous bcl-2 expression could be detected in this cell line, expression of another bcl-2 related apoptosis-suppressor protein, bcl-xL, was identified. Thus, it was suggested that bcl-2 may be functionally reduntant in this cell line [27].

3.1.2
Cell Death Following Amino Acid Starvation

The link between the onset of apoptosis and exhaustion of glutamine during batch cultures of hybridoma cells has prompted systematic studies of the role of the various nutrients used in the formulation of culture medium. Initial studies indicated that deprivation of glucose, serum, glutamine, cysteine and methionine could all, individually, induce high levels of apoptosis [9, 20, 28]. Moreover, recent studies indicate that this is not a feature of these particular nutrients alone. Deprivation, individually, of any single amino acid has been found to result in the induction of apoptosis, with particularly high levels observed following deprivation of essential amino acids [29].

Deprivation of amino acids may induce apoptosis by preventing the synthesis of molecules involved in the regulation of death. There are two ways in which this could happen. The reduction in the intra-cellular amino acid concentration would lead to an exhaustion of activated tRNA molecules resulting in the failure of translation. Alternatively, translation and transcription may be compromised by the fall in cellular ATP levels. However, not all amino acids contribute to the cellular energy requirement, yet absence of any individual amino acid results in the induction of apoptosis. Additionally, when rat hepatocytes are deprived of histidine, total protein synthesis was inhibibted at the level of peptide chain initiation [30]. Thus it would appear that it is the role of amino acids as precursors of proteins which is the most significant contributory factor to amino acid starvation induced apoptosis.

Over-expression of bcl-2 was found to offer a high degree of protection following deprivation of any single amino acid [29]. Clearly, survival of cells in the absence of essential amino acids could only be possible through the down regulation of non-essential functions and entry into a G0-like state. Indeed, recent studies demonstrate that bcl-2 can mediate cell cycle arrest as well as suppressing apoptosis. When the stress factor is removed, there is a delay of several days before cells re-enter the cell cycle. By comparison, cells transfected with a mutant bcl-2 gene which lacks the cell cycle arrest activity re-enter the cell cycle immediately after the removal of the stressing agent [31]. If wild type bcl-2 expression has a similar affect on industrially important cells, it may be necessary to utilise the mutant protein for the control of apoptosis in the bioreactor environment, thus allowing the cells to respond to improvements in the culture environment by rapidly exiting from G0, rather than remaining in an inactive and possibly a low productivity state.

Clearly, enhanced robustness in a nutrient limited environment will only be partially beneficial during the death phase of batch cultures because the cells are doomed to die anyway. However, in other culture systems, such as fed batch and high cell density perfusion systems, enhanced survival under sub-optimal nutrient concentrations may have a greater impact on the long term viability and hence productivity of the culture. In the former of these two systems, analysis of culture media components and subsequent corrective action in terms of feed constituents will be possible prior to loss of culture viability due to the increased cellular robustness. In systems such as hollow fibre perfusion reactors, improvements in reactor design can partly overcome problems associated with local nutrient limitation. Enhanced cellular robustness will clearly be important in allowing cells to tolerate such local limitations. When nutrients do become available, they can be utilised for protein biosynthesis rather than production of replacement cells.

3.1.3
Oxygen Limitation

The apparent involvement of free radical generation in the mediation of the apoptotic response has attracted a high degree of research interest. Factors such as hydrogen peroxide treatment have been found to be highly effective inducers of apoptosis, while antioxidants such as catalase and superoxide dismutase have been demonstrated to protect the cells [32]. Indeed, for a time it was thought that bcl-2 functioned as an antioxidant [33]. However, studies subsequently revealed that annoxic conditions could induce apoptosis [34], and that bcl-2 overexpression was effective at suppressing apoptosis under these conditions [5, 35].

Such studies have important consequences for large scale and intensive culture processes which suffer from oxygen limitation. Indeed, this is the key factor which limits viable cell numbers in such systems. The protective effect of bcl-2 under anoxic condtions has been shown in two industrially important cell lines [10, 36].

3.2
Susceptibility of Cells to Hydrodynamic Stress

The hydrodynamic environment of stirred tank reactors can be highly stressful to animal cells. This stems from the high energy input required to keep cells in suspension and to maintain homogeneity. Moreover, gas sparging of the culture results in high levels of cell death due to interactions with gas bubbles, specifically during bubble disengagement at the interface of the culture medium and gas headspace [37]. Whilst this knowledge has led to the development of improved bioreactor design and operation strategies that have reduced hydrodynamic stress, a compromise still exists between the need for provision of good levels of mixing and aeration, and the susceptibility of cells to hydrodynamic forces. Clearly, if such conditions result in the induction of apoptosis, the suppression of such a response should allow the operation of the bioreactor at increased agitation and sparging rates, leading to improved aeration and mass transfer with a minimal impact on culture viability. Indeed, exposure of hybridoma cells to high levels of mechanical agitation has been reported to result in significant levels of apoptotic cell death [38]. Fluorescence microscopic analysis of nuclear morphology and DNA gel electrophoresis were used to confirm the mechanism of cell death. At higher levels of hydrodynamic stress a gradual decrease in viable cell number was observed but the percentage of dead cells remained low. Flow cytometric analysis revealed features which were consistent with the induction of apoptosis (i.e. a reduction in cell size and DNA content).

Whilst the suppression of apoptosis has not been investigated under the extreme conditions of hydrodyamic stress indicated in the above study, the influence of bcl-2 on cell growth in suspension cultures of hybridoma [10] and Burkitt's Lymphoma [39] cell lines has been addressed. In the former it was found that, following transfection with control and bcl-2 expression vectors, the bcl-2 over-expressing cell line adapted to suspension culture conditions more rapidly than the control cells. In the latter example, Burkitt's Lymphoma cells which had been maintained in T-flasks exhibited poor growth and high levels of apoptosis when cultured in suspension without prior adaptation. However, over-expression of bcl-2 resulted in a significant net increase in cell number due to a reduction in the level of apoptosis.

In contrast to the above results, low level shear stress itself has been demonstrated to suppress the apoptotic response [40]. Induction of apoptosis in endothelial cells in the presence of the inducer tumour necrosis factor α or following growth factor withdrawal could be suppressed by a shear stress level of 45 dyn/cm^2. If such an effect can be demonstrated during the cultivation of industrially important animal cell lines, it may be possible to identify an optimal level of shear stress which provides both enhanced mass transfer as well as a generally applicable method for the suppression of apoptosis. Such a prospect is raised by an intriguing study of cell death in *Spodoptera frugiperda* cultures [21]. Necrosis was the predominant form of cell death in shake flask cultures of *Spodoptera frugiperda* cells. However, when cells were cultivated in a rotating wall vessel (which generates a very low turbulance environment), the level of apoptosis reached 33 % of total cells, with overall viability of 50 %. Somewhat surprisingly, culture duration was considerably enhanced despite the increase in sensitivity to apoptosis.

3.3
Anchorage Dependence

Whilst process development and scale-up for cells which grow in suspension has been relatively straightforward, large scale cultivation of anchorage dependent cell lines has posed a greater challenge. At the laboratory scale, roller bottle systems have been widely used. However, this technology does not lend itself to scale-up, is labour intensive and is highly susceptible to breakdown and contamination. The development of microcarrier technology has proved to be an important step forward. Cells grow on the surface of porous beads which can be kept in suspension in relatively simple and widely used stirred tank reactors.

The molecular basis of the anchorage dependent nature of animal cells has attracted considerable interest in recent years due to its role in tissue organisation in-vivo, and in tumour metastasis. Anchorage to the extra-cellular matrix is mediated by the cell surface receptors integrins. In addition to providing physical attachment, integrin binding also stimulates cell proliferation and suppression of apoptosis. For example, primary cultures of human breast and tracheal epithelial and human umbilical cord endotheilal cells, mink lung epithelial cells and the human prostate epithelial cell line LNCaP all undergo apoptosis when deprived of anchorage either by trypsinization or by the addition of a blocking anti-$\beta1$ integrin adhesion receptor antibody [41]. Transition from G1 to S phase of the cell cycle is determined by a group of proteins called G1 cyclin dependent kinsas (cdks). Following loss of matrix contact, it was reported that LNCaP cells exhibited an increase in the level of expression of the cdk inhibitors p21 and p27. There was also a decrease in the level of expression of cyclin D1 and E. Together, these changes result in G1 arrest of the cells following loss of matrix contact [42, 43]. In addition, cyclin D1 and cyclin E form active complexes with cdk4/6 and cdk2 which hyperphosphorylate and therefore inactivate the Retinoblastoma protein (Rb). Consequently, there was an accumulation of hypophosphorylated (activated) Rb protein which correlated with the induction of apoptosis. In contrast, cell lines with a mutated or inactive form of Rb fail to undergo apoptosis when in suspension. When Rb activity was blocked by expression of adenovirus E1a protein in LNCaP cells, apoptosis was blocked and cells were able to proliferate in suspension. Such a strategy may also be applicable to commercially important cell lines, potentially removing the need for cell growth on microcarriers or in roller bottles. Indeed, this possibility has already been demonstrated by the overexpression of bcl-2 in the industrially important cell line MDCK, which abrogated anchorage dependent growth [44].

3.4
The Requirement for Complex Culture Media

One of the main advantages of microbial cultures is the simplicity of the culture medium, which contrasts with the expensive and complex formulations required by animal cell lines. In addition to simple substrates such as amino acids, glucose, minerals and vitamins, mammalian cells also require growth factors which stimulate proliferation, and survival factors which suppress the apoptotic pathway. Serum supplementation of culture medium provides the cells with such factors. However, from a biopharmaceutical production standpoint, inclusion of serum in culture medium is highly undesirable. First, it is expensive and adds significantly to the cost of the product. Second, the protein content of serum is very high when compared to the concentration of the product. As a result, the recovery of the product from the culture medium becomes a more complex and thus a more expensive task. There are also regulatory objections surrounding its use, which necessitate screening for the presence of adventitious agents prior to use. In recent years, the development of proprietory serum-/protein-free media has made an important contribution to the development of successful large scale biopharmaceutical production processes, although the cost of such media is high. Further improvements in protein-free media may be possible through the application of our understanding of the role of serum components in the induction of apoptosis. The inclusion of chemicals which can specifically suppress the induction of apoptosis should substantially improve the performance of such media. This possibility is clearly indicated by a recent study in which the fall in viability which accompanied the reduction in the serum concentration from 10% to 1% in Sp2/0 hybridoma cultures could be blocked by supplementation of culture medium with IL-6 [45]. Other chemicals which could potentially play such a role are listed in Table 2.

One particularly important class of anti-apoptosis factors are the inhibitors of Interleukin 1β-converting enzyme (ICE) proteases (now referred to as caspases) [46]. This family of enzymes (which comprises at least ten members) plays a key role in transducing the death signal from the initial interactions with the inducing agent and also appears to be involved in the effector phase of death. A recent study demonstrated that apoptosis induced by camptothecin treatment (which induces apoptosis in S-phase cells by inhibiting topisomerase 1) could be prevented by the incubation of cells with the cell permeable ICE inhibitor benzyloxycarbony-Val-Ala-Asp (O-methyl)-fluoromethylketone (Z-VAD-fmk) [47]. Addition of Z-VAD-fmk to cultures also completely inhibited the anti-Fas antibody induced apoptosis in SKW 6.4 cells [48]. In the human hepatoma cell line Hep-3, Z-VAD-fmk prevented TGF-β-induced apoptosis [49]. Clearly this particular inhibitor is highly effective at suppressing apoptosis induced by a wide range of agents in a wide range of cell types. The effectiveness of such agents during the cultuvation of industrially important cell lines requires urgent investigation.

Expression of anti-apoptosis genes may provide an alternative method of improving cell growth in simple media. For example, a Burkitt's Lymphoma cell line transfected with bcl-2 grew much better on serum-free medium without the need for prior adaptation [39]. However, transfection of a hybridoma cell line with the cDNA for IL-6 did not enhance viability to the same extent as addition of exogenous IL-6 [45].

3.5
Low Cell Densities

Cell lines such as hybridomas will typically reach a maximum cell number of around 1×10^6 to 2×10^6 ml^{-1} in suspension batch cultures. Increased production can be achieved either through nutrient feeding or by means of process intensification. Because of the high nutrient and oxygen demand of mammalian cells, conditions in intensive culture systems are particularly stressful, as discussed earlier. As a result, such systems are characterised by a particularly high level of cell death. For example, studies have demonstrated that apoptosis accounts for much of the cell death during the cultivation of hybridoma cells in perfusion cultures utilising a spin filter to retain cells in the bioreactor [50]. Moreover, it was reported that the overexpression of the anti-apoptotic gene bcl-2 protected cells from the prevailing stress, leading to a doubling in viable cell number and an increase in viability. Studies in fixed bed and hollow fibre bioreactors indicated that the enhanced robustness of the bcl-2 transfected cells resulted in a two-fold increase in antibody productivity over the control cell line [50]. In both systems, it proved difficult to measure viable cell number and viability directly. However, an indirect method involving the fluorimetric estimation of free DNA in the culture medium was developed and used with the fixed bed system. This revealed a higher free DNA content in the culture medium from the production run using the control cell lines, thus indicating a higher rate of cell death.

3.6
Low Cell Productivity

The quantity of product secreted by a mammalian cell is very low (of the order of pg/ml). Thus, a number of studies have centred on devising strategies which optimise specific productivity. An important component of this work concerns the study of the relationship between cellular proliferation and protein productivity. The nature and basis of this relationship is a highly complex one which is dependent on a range of factors including culture conditions, cell type, the cell cycle and the expression vector system utilised. However, generally cell lines can be classed into two groups: those that exhibit a positive correlation between growth rate and productivity, and those which exhibit a negative correlation. In this context, the study of hybridoma cells under conditions of growth arrest is of particular interest. This cell type exhibits a negative correlation, i.e. as the growth rate decreases the specific antibody productivity of the cells increases dramatically. This was illustrated by the finding that cell cycle arrest by thymidine treatment resulted in a substantial increase in specific antibody productivity [51, 52]. However, such a state cannot be maintained for very long due to a reduction in culture viability. It is now clear that this is due to the induction of apoptosis, which can be suppressed by the expression of bcl-2 [9, 10, 39]. As discussed earlier, expression of the adenovirus E1B-19Kd gene in NS0 cells has been demonstrated to provide protection from apoptosis following nutrient deprivation, resulting in a significant reduction in the rate of cell death during the death phase of batch cultures [26]. However, under these conditions there was no improvement in antibody titre over control cultures. When cells were cultured under growth arrest conditions by treatment with thymidine, hyperosmotic conditions or in the presence of OptiMab, culture viability for the E1B-19 Kd overexpressing cell line remained over 80% over a 6 day period, as compared to less than 25% for the control cell line. In the case of OptiMab treated cultures, this enhancement in viability led to a 350% increase in antibody producivity for E1B-19Kd transfected cells, compared to only a 75% increase for control cells.

Simultaneous cell cycle arrest and suppression of apoptosis has been achieved by transfection of F-MEL cells with the c-jun antisense gene under the control of the glucocorticoid-inducible MMTV promoter [53]. Treatment with dexamethasone resulted in cell cycle arrest with viability remaining above 92% over a 16 day period. When the cells were exposed to simultaneous growth arrest and serum deprivation, viability had only fallen to around 86% after ten days. When the experiment was repeated with wild-type F-MEL cells, the viability was reduced to 50% over the same period, with the first indication of apoptosis after only 48 h.

4
Programmed Cell Death and Bioprocess Technology: Future Prospects

It should be apparent from the above discussion that apoptosis research has now been established as an important component of animal cell biotechnology. With competition between biotechnology/pharmaceutical companies now beginning to increase, it is only a matter of time before apoptosis-resistant cell lines make the transition from the research laboratory to the production plant. In this section we highlight areas of biotechnology which will benefit from apoptosis research in the next few years.

4.1
Selection of Apoptosis-Resistant Cell Lines

Variation in susceptibility to apoptosis between completely unrelated cell lines is clearly evident. Therefore the initial decision to adopt a cell line for a particular production process should take into consideration the susceptibility to apoptosis. Further research is also required to assess the clonal variation in susceptibility to apoptosis within a particular cell line, and to relate this to the level of endogenous anti-apoptotic gene expression.

4.2
Other Targets for the Genetic Suppression of Apoptosis

The extent of the protection offered by the anti-apoptosis genes investigated thus far is dependent upon the level of expression of the gene in question. Thus, expression of high levels of such genes would be expected to impinge on the expression of the recombinant protein of interest. Although only one study has revealed any significant reduction in antibody titre over control cultures following transfection with an anti-apoptosis gene [26], it would be expected that in all cases productivity could be further enhanced by use of a protein which can provide high levels of protection whilst expressed at a relatively low level.

4.3
Tissue Engineering

Tissue engineering harnesses both the engineering and biological disciplines to provide functional substitutes for damaged tissues and organs [54]. Important areas include the use of skin cultures for the treatment of burns patients, the development of bioartificial livers for the treatment of patients with acute or chronic liver failure, and ex-vivo expansion of hemopoietic stem cells for therapeutic applications such as bone marrow transplantation and hosts for gene therapy trials.

Development of successful tissue/organ replacements, such as those described above, will draw heavily on our knowledge of the role of apoptosis in organ development during embryogensis and in normal cell turnover and organisation of tissues in the fully developed organism. By way of example, one can consider the role of hormones in the control of haemopoeitic stem cell growth and differentiation. Initial studies demonstrated that ex-vivo stem cell cultivation required bone marrow stroma, which is comprised of the various cell types that constitute the environment of stem cells in bone marrow. These cells play an important role in stem cell proliferation, differentiation and survival by providing a range of bound and soluble hormones. Amongst these hormones are two which have been specifically implicated in the suppression of apoptosis: stem cell factor, which suppresses apoptosis in early progenitor cells, and IL-3 which suppresses apoptosis in committed progenitor cells. The use of stroma for the production of stem cells as therapeutics is far from ideal. Addition of hormones to suspension cultures of stem cells does facilitate cultivation without the need for stroma (reviewed in [55]). The cost and complexity of such an approach may be reduced by the addition of general apoptosis-suppressing agents, such as ICE inhibitors, which would clearly simplify the hormone requirements of the cells. Furthermore, because apoptosis plays such a central role in the immune system, it is quite likely that these cells are highly susceptible to apoptosis induced by factors such as nutrient limitation. Thus, systematic studies into the role of the culture environment on apoptosis during stem cell expansion and the role of apoptosis-inhibiting agents on the success of such cultures is required.

4.4
Programmed Cell Death and Single-Celled Organisms

For many years now it has been argued that the induction of truly altruistic cell death is only possible in multicellular organisms in which death of the individual cell benefits the organism as a whole. Because the genetic material of the cells which undergo apoptosis is propagated by germ line cells, the genes which encode cell death can be passed on from generation to generation. However, this view is increasingly being challenged by reports of altruistic cell death in microbial cells. For example, studies on *Schizosaccharomyces pombe* suggest that the apoptotic pathway may also be conserved in this unicellular organism. When the bcl-2 related protein BAK (which induces apoptosis in mammalian cells) was overexpressed in *S. pombe*, morphological features associated with apoptosis were observed [56]. Moreover, when cells were co-transfected with BAK and bcl-XL, cell death was suppressed. Programmed cell death pathways also exist in bacteria. There have been reports of bacteriophage-induced suicide of bacterial cells, which is believed to prevent the spread of the infection to the remainder of the colony [57].

Bacteria and yeast are important hosts for the production of a wide range of fine chemicals. Identification and regulation of programmed cell death during such a process may well widen the impact of programmed cell death research on biotechnology.

5
References

1. Drews J (1993) Biotechnol 11:S16–S20
2. Ohyama H, Yamada T, Watanabe I (1981) Rad Res 85:333–339
3. Thomas N, Bell PA (1981) Mol Cellul Endocrinol 22:71–84
4. Sanderson CJ (1982) Morphological aspects of lymphocyte mediated cytotoxicity. In: Clark WR, Golstein P (eds) Mechanisms of cell mediate cytotoxity. Plenum, New York, pp 3–21
5. Jacobson MD, Raff M C (1995) Nature 374 814–816
6. Oberhammer F, Wilson JW, Dive C, Morris ID, Hickman JA, Wakeling AE, Walker PR, Sikorska M (1993) EMBOJ 12:3679–3684
7. Wyllie AH (1980) Nature 284:555–556
8. Piacentini M, Fesus L, Farrace MG, Ghibelli L, Piredda L, Melino G (1991) Euro J Cell Biol 54:246–254
9. Singh RP, Al-Rubeai M, Gregory CD, Emery AN (1994) Biotechnol Bioeng 44:720–726
10. Simpson N, Milner AN, Al-Rubeai M (1997) Biotechnol Bioeng 54:1–16
11. Steck KD, Mcdonnell TJ, Elnaggar AK (1995) Cytometry 20:154–161
12. Ormerod MG, Sun XM, Snowden RT, Davies R, Fearnhead H, Cohen GM (1993) Cytometry 14:595–602
13. Elstein KH, Zucker RM (1994) Exp Cell Res 211:322–331
14. Elstein KH, Thomas DJ, Zucker RM (1995) Cytometry 21:170–176
15. Nicoletti I, Migliorati G, Pagliacci MC, Grignani F, Riccardi C (1991) J Immunol Methods 139:271–279
16. Fadok VA, Voelker DR, Campbell PA, Cohen JJ, Bratton DL, Henson PM (1992) J Immunol 148:2207–2216
17. Koopman G, Reutelingsperger CPM, Kuijten GAM, Keehnen RMJ, Pals ST, Vanoers MHJ (1994) Blood 8:1415–1420
18. Al-Rubeai M, Mills D, Emery AN (1990) Cytotechnol 4:13–28
19. Franek F, Dolnikova J (1991) FEBS Letters 248:285–287
20. Mercille S, Massie B (1994) Biotechnol Bioeng 44:1140–1154
21. Cowger NL, O'Conner KC (1997) Poster 8P20 presented at the 15th meeting of the European Society for Animal Cell Technology at Tours, Val-de-Loire, France
22. Moore A, Donahue CJ, Hooley J, Stocks DL, Bauer KD, Mather JP (1995) Cytotechnol 17:1–11
23. Chatzisavido N, Fenge C, Haggstrom L (1997) Poster number 8P16 presented at the 15th meeting of the European Society for Animal Cell Technology at Tours, Val-De-Loire, France
24. Itoh Y, Ueda H, Suzuki E (1995) Biotechnol Bioeng 48:118–122
25. Suzuki E, Terada S, Ueda H, Fujita T, Komatsu T, Takayama S, Reed JC (1997) Cytotechnol 23:55–59
26. Mercille S, Gervais C, Paquette D, Massie B (1997) Presented at the 15th meeting of the European Society for Animal Cell Technology at Tours, Val-De-Loire, France
27. Murray K, Ang CE, Gull K, Hickman JA, Dickson AJ (1996) Biotechnol Bioeng 55:298–304
28. Perreault J, Lemieux R (1994) Cytotechnol 13:99–105
29. Simpson NH, Singh RP, Perani A, Goldenzon C, Al-Rubeai M (1998) Biotechnol Bioeng (to appear)
30. Kimball SR, Yancisin M, Horetsky RL, Jefferson LJ (1996) Int J Biochem Cell Biol 28:285–294
31. Huang DCS, Oreilly LA, Strasser A, Cory S (1997) EMBO J 16:4628–4638
32. Tilly JL, Tilly KI (1995) Endocrinol 136:242–252
33. Hockenbery DM, Oltvai ZN, Yin XM, Milliman CL, Korsmyer SJ (1993) Cell 75:241–251
34. Mercille S, Massie B (1994) Cytotechnol 15:117–128
35. Shimizu S, Eguchi Y, Kosaka H, Kamiike W, Matsuda H, Tsujimoto Y (1995) Nature 374 811–813

36. Singh RP, Finka G, Emery AN, Al-Rubeai M (1997) Cytotechnol 23:87–93
37. Handa-Corrigan A, Emery AN, Spier RE (1989) Enzyme Microbial Technol 11:230–235
38. Al-Rubeai M, Singh RP, Goldman MH, Emery AN (1994) Biotechnol Bioeng 45:463–472
39 Singh RP,Emery AN, Al-Rubeai M (1996) Biotechnol Bioeng 52:166–175
40. Dimmeler S, Haendeler J, Rippmann V, Nehls M, Zeiher AM (1996) FEBS Letters 399:71–74
41. Day ML, Foster RG, Day KC, Zhao X, Humphrey P, Swanson P, Postigo AA, Zhang SH, Dean D (1997) J Biol Chem 272:8125–8128
42. Zhu X, Ohtsubo M, Bohmer RM, Roberts JM, Assoian RK (1996) J Cell Biol 133:391–403
43. Fang F, Orend G, Watanabe N, Hunter T, Ruoslahti E (1996) Science 271:499–503
44. Frisch SM, Francis H (1994) J Cell Biol 124:619–626
45. Chung JD, Zabel C, Sinskey AJ, Stephanopoulos G (1995) Biotechnol Bioeng 55:439–446
46. Alnemri ES, Livingstone DJ, Nicholson DW, Salvesen G, Thornberry NA, Wong WW, Yuan J (1996) Cell 87:171
47. Shimizu T, Pommier Y (1997) Leukemia 11:1238–1244
48. Sillence DJ, Allan D (1997) Biochemical Journal 324 29–32
49. Chen RH, Chang TY (1997) Cell Growth Differ 8:821–827
50. Perani A, Singh RP, Fassnacht D, Portner R, Al-Rubeai M (1997) Manuscript in preparation
51. Al-Rubeai M, Emery AN (1990) J Biotechnol 16:67–86
52. Al-Rubeai M, Emery AN, Chalder S, Jan DC (1992) Cytotechnol 9:85–97
53. Kim YH, Iida T, Fujita T, Terada S, Kitayama A, Ueda H, Suzuki E (1997) Presented at the 15th meeting of the European Society for Animal Cell Technology at Tours, Val-de-Loire, France
54. Langer R, Vacanti JP (1993) Science 260:920–926
55. McAdams TA, Sandstrom CE, Miller WM, Bender JG, Papoutsakis ET (1995) Cytotechnol 18:133–146
56. Ink B, Zornig M, Baum B, Hajibagheri N, James C, Chittenden T, Evan G (1997) Mol Cellul Biol 17:2468–2474
57. Shub DA (1994) Curr Biol 4:55–556
58. Boise LH, Gonzalezgarcia M, Postema CE, Ding LY, Lindsten T, Turka LA, Mao XH, Nunez G, Thompson CB (1993) Cell 74:597–608
59. Nagat S (1997) Cell 88:355–365
60. Tsujimoto Y, Cossman J, Jaff E, Croce CM (1985) Science 228:1440–1443
61. Clem RJ, Miller LK (1994) Mol Cell Biol 14:5212–5222
62. Henderson S, Huen D, Rowe M, Dawson C, Johnson G, Rickinson AB (1990) Proc Natl Acad Sci USA 90:8479–8483
63. Peitsch MC, Polzar B, Stephan H, Crompton T, Macdonald HR, Mannherz HG, Tschopp J (1993) EMBO J 12:371–377
64. Barry MA, Eastman A (1993) Arch Biochem Biophys 300:440–450
65. Gaido ML, Cidlowski JA (1991) J Biol Chem 266:580–585
66. Alnemri ES, Livingstone DJ, Nicholson DW, Salvesen G, Thornberry NA, Wong WW, Yuan J (1996) Cell 87:171
67. Piacentini M, Fesus L, Farrace MG, Ghibelli L, Piredda L, Melino G (1991) Euro J Cell Biol 54:246–254
68. Ferrari G, Yan CYI, Greene LA (1995) J Neurosc 15:2857–2866
69. Okada N, Koizumi S (1995) J Biol Chem 270:16,464–16,469
70. Catchpoole DR, Stewart BW (1994) Anticancer Res 14:853–856
71. Traganos F, Kapuscinski J, Gong JP, Ardelt B, Darzynkiewicz RJ, Darzynkiewicz Z (1993) Cancer Res 53:4613–4618
72. Squier MKT, Miller AK, Malkinson AM, Cohen JJ (1994) J Cellul Physiol 159:229–237
73. Rami A, Krieglstein J (1993) Brain Res 609:67–70
74. Sarin A, Clerici M, Blatt SP, Hendrix CW, Shearer GM, Henkart PA (1994) J Immunol 153:862–872

75. Wilson KP, Black JAF, Thomson JA, Kim EE, Griffith JP, Navia MA, Murcko MA, Chambers SP, Aldape RA, Raybuck SA, Livingston DJ (1994) Nature 370:270–275
76. Walker NPC, Talanian RV, Brady KD, Dang LC, Bump NJ, Ferenz CR, Franklin S, Ghayur T, Hackett MC, Hammill LD, Herzog L, Hugunin M, Houy W, Mankovich JA, Mcguiness L, Orlewicz E, Paskind M, Pratt CA, Reis P, Summani A, Terranova M, Welch JP, Xiong L, Moller A, Tracey D, Kamen R, Wong W (1994) Cell 78:343–352
77. Thornberry NA, Peterson EP, Zhao JJ, Howard AD, Griffin PR, Chapman KT (1994) Biochemistry 33:3934–3940

Received February 1998

Author Index Volume 1–62

Author Index Vols. 1–50 see Vol. 50

Subject Index

Springer
and the
environment

At Springer we firmly believe that an international science publisher has a special obligation to the environment, and our corporate policies consistently reflect this conviction.

We also expect our business partners – paper mills, printers, packaging manufacturers, etc. – to commit themselves to using materials and production processes that do not harm the environment. The paper in this book is made from low- or no-chlorine pulp and is acid free, in conformance with international standards for paper permanency.

Springer

Printing: Saladruck, Berlin
Binding: Buchbinderei Lüderitz & Bauer, Berlin